Chemical Fingerprints of Medicinal Plants - HPTLC Profiling

Authored By

Acharya Balkrishna
University of Patanjali
Patanjali Research Foundation Trust
Haridwar
India

Chemical Fingerprints of Medicinal Plants - HPTLC Profiling

Author: Acharya Balkrishna

ISBN (Online): 978-981-5196-41-2

ISBN (Print): 978-981-5196-42-9

ISBN (Paperback): 978-981-5196-43-6

First published in 2023.

need for a court order if at any point you breach any terms of this License Agreement. In no event will any delay or failure by Bentham Science Publishers in enforcing your compliance with this License Agreement constitute a waiver of any of its rights.

3. You acknowledge that you have read this License Agreement, and agree to be bound by its terms and conditions. To the extent that any other terms and conditions presented on any website of Bentham Science Publishers conflict with, or are inconsistent with, the terms and conditions set out in this License Agreement, you acknowledge that the terms and conditions set out in this License Agreement shall prevail.

Bentham Science Publishers Pte. Ltd.
80 Robinson Road #02-00
Singapore 068898
Singapore
Email: subscriptions@benthamscience.net

BENTHAM
SCIENCE

CONTENTS

PREFACE

Herbs have been used for medicinal purposes since ancient time. A large number of populations in the world rely on herbs and herbal extracts for different ailments. The Indian traditional medicine system – "Ayurveda" also recognizes the importance of the herbs and their authenticity.

In the words of Sage Acharya Charaka,

Tasyāpīyaṃparīkṣā- Idamevaṃprakṛtyevaṃguṇamevaṃprabhāvamsmindeśe

jātamasminnṛtāvevaṃ gṛhītamevaṃnihitamevamupaskṛtamanayā ca mātrayā

CarakaSaṃhitāVimānasthāna 8:87

"Meaning - for selection of herbs Organoleptic, physical and chemical characteristics, habitat, season, method of collection, method of preservation, processing, and dosage should be considered"

With the increasing reliance of the worldwide population on herbs, the selection of the right quality of herbs, identification, and authentication are the need of time. In the modern day, different techniques like taxonomy, biological and chemical fingerprinting are in use to validate the procedure for the selection of the right quality herb.

In the present book, we have developed and compiled the High-Performance Thin Layer Chromatography (HPTLC) fingerprints of some of the commonly used medicinal plants. HPTLC fingerprint provides a scientific approach for the identification and authentication of herbal material as it provides a unique phytochemical distribution pattern which can be seen in the form of bands and can further be used for the purpose of qualitative and quantitative checks.

I am sure that this book will be immensely helpful as a reference book for the scientific community working in this area throughout the nation and world over.

Acharya Balkrishna
University of Patanjali
Patanjali Research Foundation Trust
Haridwar
India

CHAPTER 1

Introduction to High-Performance Thin Layer Chromatography (HPTLC)

Abstract: This chapter discusses chromatography, classification of chromatography, thin layer chromatography (TLC) and how high-performance thin-layer chromatography (HPTLC) is an advanced version of TLC. In brief, this chapter also discusses the advantages and the areas where the HPTLC technique can be used.

Keywords: Chromatography, Densitometry, HPTLC, TLC, TLC fingerprint, UV 254 nm, UV 366 nm, Visualization.

INTRODUCTION

Chromatography is a technique or a process that is used to separate compounds from a mixture. It consists of two different phases: one is stagnant which is the stationary phase. The other is the mobile phase which moves through the stationary phase and separates out the components on the basis of their interactions with the mobile and stationary phases. The stationary phase can be solid (silica gel) or liquid, and mobile phase can be in liquid, gas, and supercritical form.

Chromatography can be classified on the basis of the mobile phase into liquid chromatography and gas chromatography. It is also classified on the basis of separation mechanism: ion exchange, size exclusive and adsorption, and can also be categorised by the shape of the bed into column chromatography and planer chromatography (TLC).

THIN LAYER CHROMATOGRAPHY (TLC)

TLC is a chromatographic technique that is used to elute out compounds on a thin layer of stationary phase generally prepared from silica gel.

For performing TLC/ HPTLC analysis, first, we dissolved the sample in a suitable solvent. The prepared sample is then applied on the TLC plate either manually or by autosampler (in HPTLC) in the form of a band or a spot. This TLC plate is then transferred to the chamber containing the mobile phase for development.

Acharya Balkrishna

As the mobile phase passes through the stationary phase, it allows the compounds to be separated on the basis of their polarity or affinity with both phases. As the mobile phase migrates toward the marked solvent front, the TLC plate is removed and dried. The developed TLC plate is then visualized under UV 254 nm, 366 nm, and white light with or without derivatization on the basis of the molecule to be detected [5].

HPTLC is the mechanized and advanced form of TLC as it allows the application of samples through the automatic sampler, with complete documentation of chromatographic data using software, and also with an auto scanner which helps in the identification of compounds. It provides both qualitative and quantitative analysis which is not possible in manual TLC methods.

The use of HPTLC allows proper separation with high resolution, repeatability, analysis of multiple samples together, visual inspection of chromatograms, and fast delivery of results.

TLC/HPTLC can successfully be hyphenated with other analytical techniques like High-performance liquid chromatography (HPLC), Mass spectroscopy (MS) and Fourier transform infrared (FTIR), which can give new dimensions to the research works [5].

Advantages of TLC/HPTLC

TLC offers several advantages:

- Visual representation and documentation of results.
- Parallel analysis of samples for fast screening.
- Flexibility.
- Detection of multiple samples.

Visual Representation and Documentation of Results

TLC provides the option to visualize the chromatograms in the form of the TLC plate image. TLC plate can be visualized under UV 254 nm and 366 nm for UV active compounds and some compounds can also be documented after derivatization using different spraying reagents on the basis of the class of compounds. Compounds are visualized in the form of bands and provide a fingerprint of that particular sample Fig. (**1**)

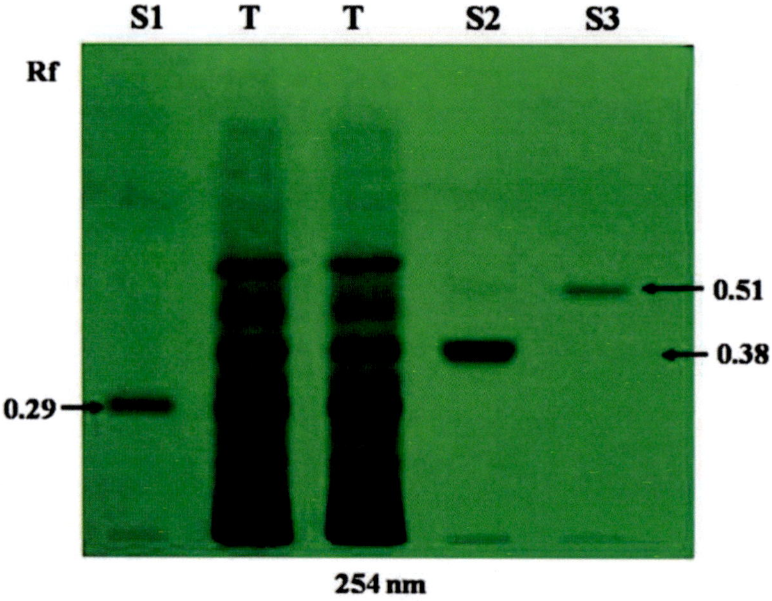

Fig. (1). HPTLC fingerprint of *Withania somnifera* (L.) Dunal at 254 nm.
Mobile phase (MP): Toluene: ethyl acetate: formic acid (5: 5: 1 v/v/v). S1: Witheferin A (Rf 0.28), T: *Withania somnifera* (L.) Dunal, S2: Withanolide A (Rf 0.20), S3: Withanolide B

Parallel Analysis of Samples for Fast Screening

This technique enables parallel analysis of multiple samples. In one TLC plate, we can apply approximately 18 different samples or same samples of different batches. TLC plate is developed under same chromatographic conditions and visualized, and it provides fast screening of multiple samples at one time Fig. (**2**).

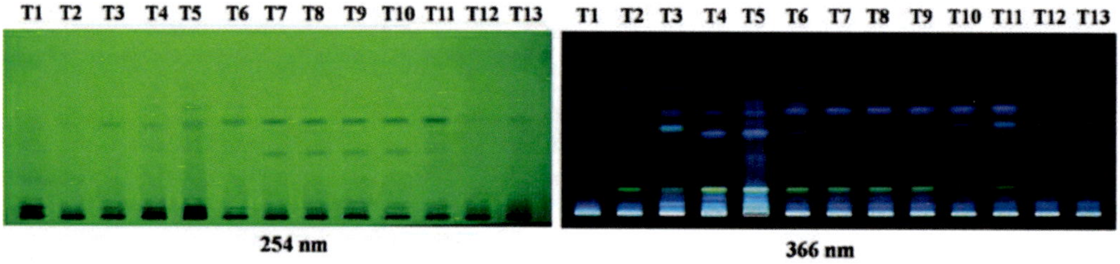

Fig. (2). Parallel analysis of multiple samples. MP: Ethyl acetate: toluene: formic acid (5: 5: 1 v/v/v).
A: 254 nm, B: 366 nm.

Flexibility

TLC is a flexible technique where different combinations of mobile phases can be used on the basis of the molecules to be separated.

Detection of Multiple Samples

HPTLC provides the advantage of multiple detections using the same TLC plate. It means different types of compounds can be analyzed in one TLC plate on the basis of their detection wavelength. Developed TLC plate is visualized under UV 245 nm, 366 nm, and white light before and after derivatization. Similarly for quantitative analysis, the same plate can be scanned on different wavelengths on the basis of the molecule to be analyzed Fig. (**3**).

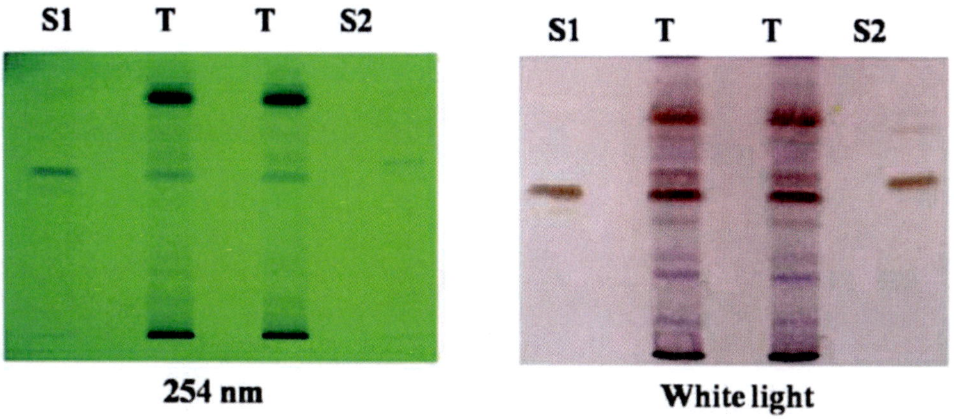

Fig. (3). Multiple detection using same plate at different wavelength. HPTLC fingerprint of *Zingiber officinale* Roscoe. MP: Toluene: ethyl acetate: diethyl ether (12: 6: 2 v/v/v) A: at 254 nm, B: Under white light after derivatization (anisaldehyde sulfuric acid reagent). S1: 6 Gingerol, T: *Zingiber officinale* Roscoe, S2: 8 Gingerol.

Application Fields

HPTLC technique is very useful in different working areas such as phytochemistry, pharmaceutical, food industry, forensic science, biotechnology, and cosmetics, and chemical and environmental sciences. HPTLC analysis in different areas can be performed to check the quality of material, batch-to-batch consistency, uniformity of content, to check adulteration, and to generate fingerprinting Figs. (**4** and **5**).

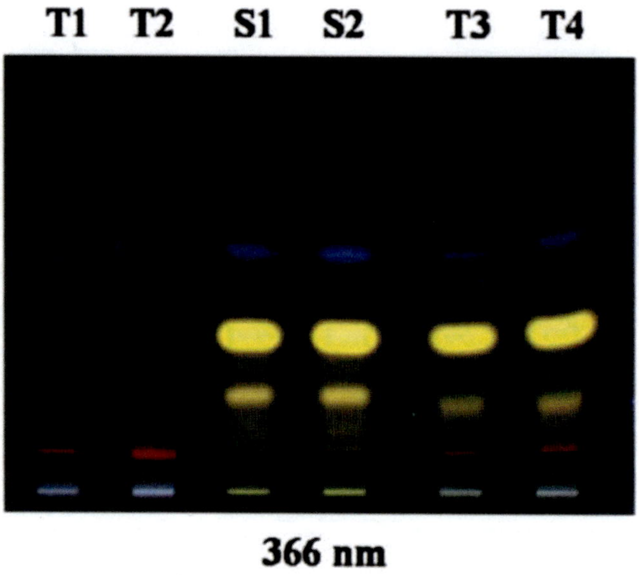

Fig. (4). HPTLC fingerprint of mustard oil and adulterant at 366 nm, MP: Hexane: ethyl acetate (16: 4 v/v). T1 and T2: Mustard oil, S1, and S2: Argemone, T3, and T4: Mustard oil spiked with argemone.

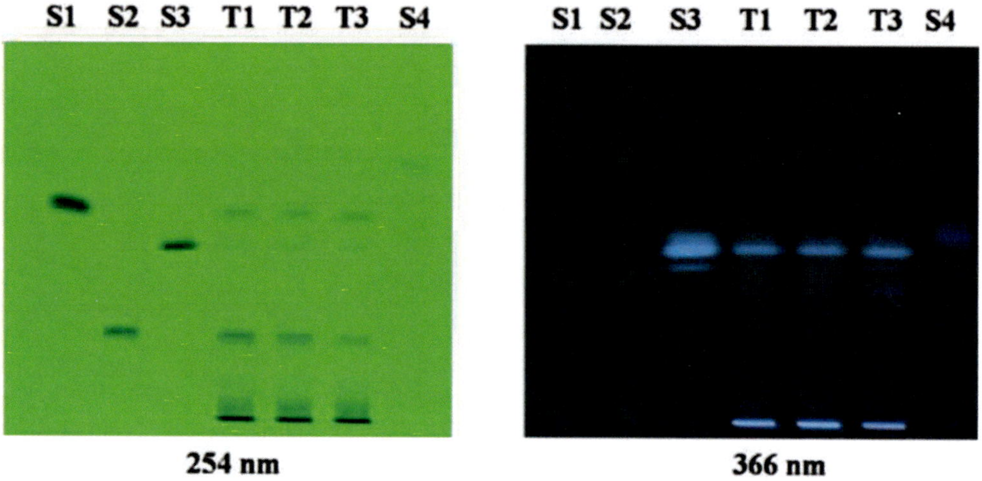

Fig. (5). HPTLC fingerprint of Chyawanprash. MP: Ethyl acetate: toluene: formic acid (18: 20: 1 v/v/v), S1: Cinnamic acid, S2: Gallic acid, S3: Piperine, T1, T2 and T3: Chyawanprash, S4: Eugenol.

CONCLUDING REMARKS

HPTLC is a versatile, reliable and quick method, which is applicable in various fields like food, forensics, biotechnology, cosmetics, pharmaceutical, chemical and environmental. For herbs, HPTLC/ TLC technique can be used for identification and authentication of herbal plants and products derived from herbal sources. This technique offers many advantages over other techniques. Different samples can be analyzed at the same time, which can be very useful and a time-saving technique to check the mixing of impurities in the materials. It also provides qualitative as well as quantitative results.

CHAPTER 2

High-Performance Thin Layer Chromatography Instrumentation

Abstract: HPTLC is one type of planar chromatography and the most advanced form of instrumental TLC. HPTLC equipment involves an autosampler for the application of a specific amount of sample on the plate, after spotting of sample the plate is developed in a TLC development chamber after development the plate is dried and visualized using visualizer. This chapter briefly discusses the equipment involved in the development of TLC.

Keywords: ATS 4, AMD 2, CAMAG, Densitometry, TLC.

INTRODUCTION

Sample Applicator

The quality of the chromatographic results depends on sample application. For quality results, samples must be positioned precisely, sample volume must be controlled and the layer must not be damaged.

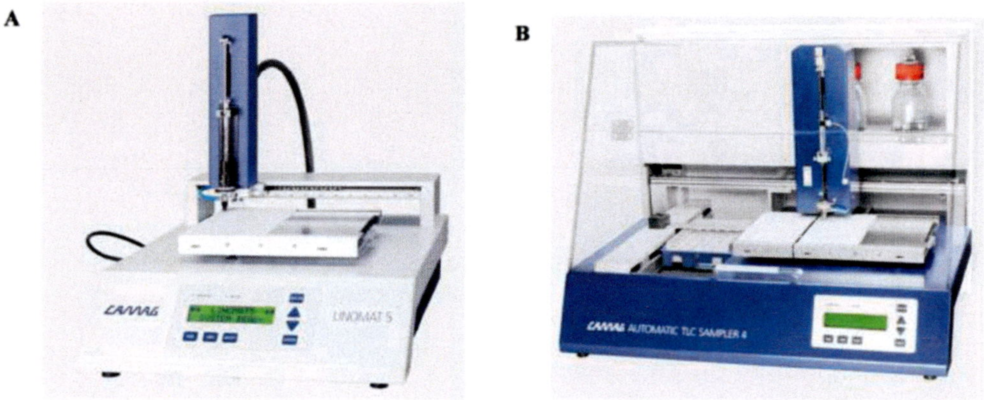

Fig. (6). Instrument for sample application. (**a**) CAMAG Linomate 5, (**b**) CAMAG Automatic TLC Sampler 4.

Acharya Balkrishna

In HPTLC, different instruments are available for semiautomatic and automatic sample application. The most widely used sample applicator featuring the spray-on technique is CAMAG Linomat 5 Fig. (**6A**) and CAMAG Automatic TLC Sampler 4 Fig. (**6B**). All application parameters are computer-controlled and programmed by CAMAG winCATS software. CAMAG Linomat 5 is semiautomatic in which the sample is loaded into the syringe manually. CAMAG ATS 4 is an automatic sample applicator as it allows the application of samples by both contact spotting and spray-on techniques [5].

TLC Developer

Chromatogram development is the key element of the process. There are different types of chambers that can be used for development: flat-bottom chamber Fig. (**7A**), twin-trough chamber Fig. (**7B**), automatic development chamber Fig. (**7C**) and automatic multiple development chamber Fig. (**7D**).

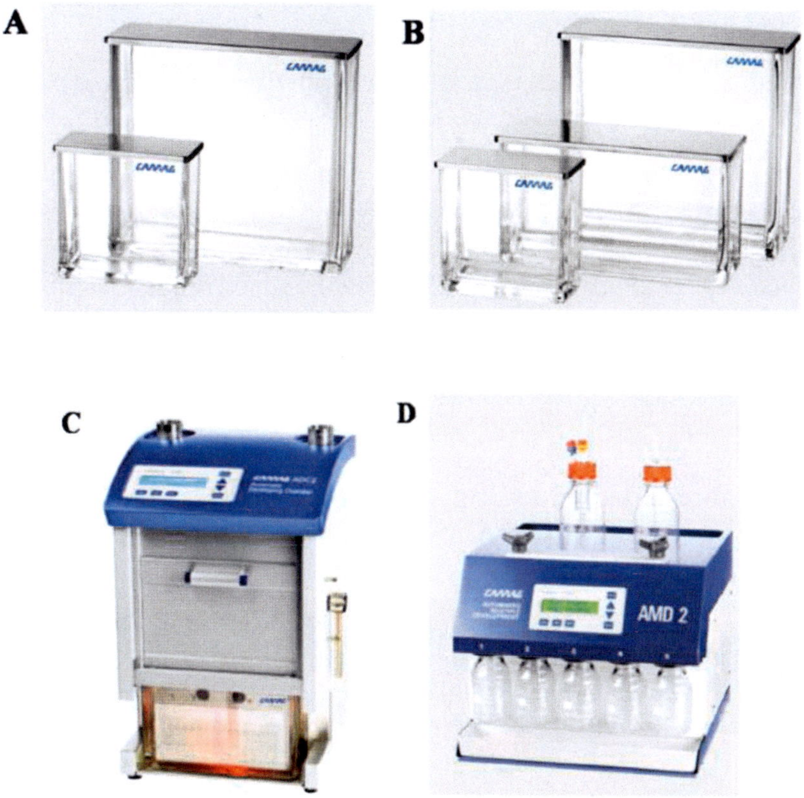

Fig. (7). Different types of development chambers. (**a**) Flat-bottom chamber, (**b**) Twin-trough chamber, (**c**) Automatic development chamber (ADC 2), (**d**) Automatic multiple development chamber (AMD 2).

A flat-bottom chamber permits the development of the plate under conditions of partial or complete saturation of the gas phase with solvent vapours.

Twin-trough chamber (TTC) is economic as well as flexible. A TTC is charged with a solvent to reach an initial level of 5 mm or 10 mm in the trough that receives the plate. The same amount is used in the rare trough if chamber saturation is established.

Automatic Development Chamber 2 (ADC 2,) is a fully automatic development chamber that generates reproducible results independent of handling errors. After placing the TLC plate into the device, the layer is automatically brought in contact with the mobile phase and development happens [4].

In automatic multiple development chamber (AMD 2), gradient mobile phase can be used. It is a fully automatic technique.

DERIVATIZER

Derivatization is performed to visualize fingerprints. This can be achieved through the gas phase and reagents. Different instruments used for derivatization in HPTLC/TLC are described below.

The glass reagent sprayer Fig. (**8A**) comes with a rubber pump but also be operated from a compressed air or nitrogen supply. CAMAG TLC sprayer Fig. (**8B**) is used for manual spraying of derivatization reagents onto TLC plates. CAMAG TLC spray cabinet Fig. (**8C**), ensures the complete removal of reagent mist while manually spraying onto the TLC plate. CAMAG Chromatogram immersion device Fig. (**8D**) is used for automatic dipping. HPTLC/TLC plates are immersed with uniform vertical speed into derivatization reagents. CAMAG TLC plate heater Fig. (**8E**), is designed for heating HPTLC/TLC plates at required temperature before and after derivatization [4].

VISUALIZER

TLC offers the advantage of visualizing chromatographic results directly in the form of fingerprints or chromatograms using UV light, at 254 nm and 366 nm and white light. CAMAG TLC Visualizer 2 Fig. (**9**) is used for imaging and documentation of TLC fingerprints (V) (https://www.camag.com).

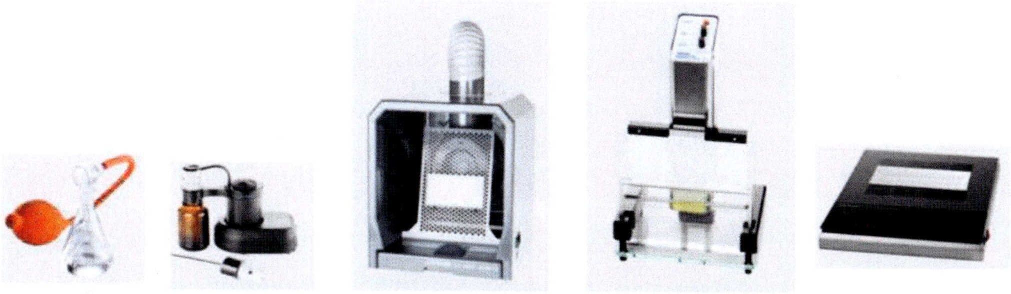

A B C D E

Fig. (8). Different types of instruments used for derivatization. (**a**) Glass reagent sprayer, (**b**) CAMAG TLC sprayer, (**c**) CAMAG TLC spray cabinet, (**d**) CAMAG Chromatogram immersion device, (**e**) CAMAG TLC plate heater.

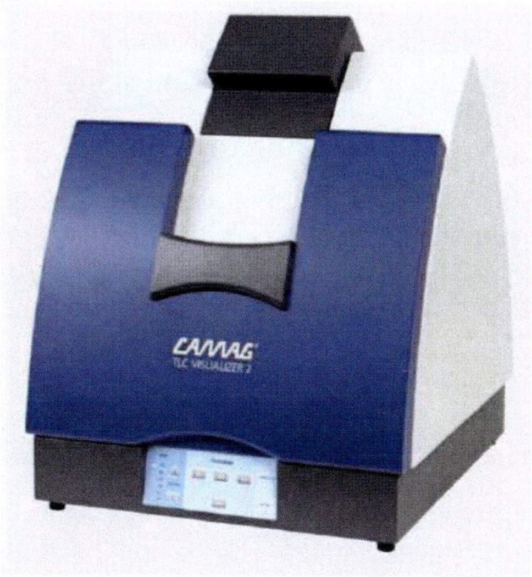

Fig. (9). CAMAG TLC Visualizer for imaging and documentation.

DENSITOMETRY

Densitometry is measuring and quantifying a thin layer chromatography. CAMAG TLC Scanner 4 Fig. (**10**) is used for densitometry analysis.

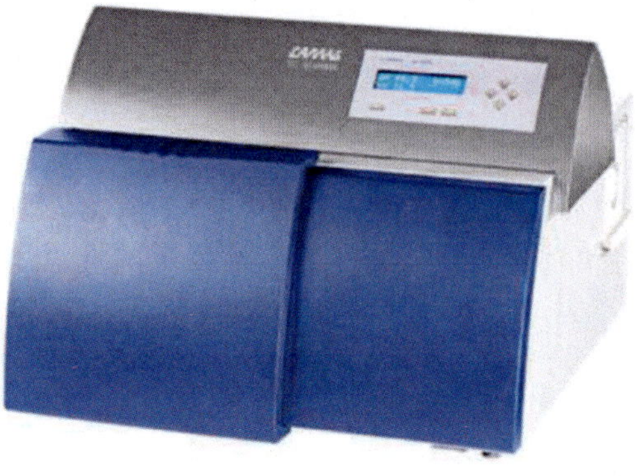

Fig. (10). CAMAG TLC Scanner 4 for densitometric evaluation of TLC chromatograms.
Note: HPTLC equipment images have been sourced from https://www.camag.com.

CONCLUDING REMARKS

HPTLC instruments make this technique the most flexible and versatile. This technique allows the application of up to 66 samples from standard sample vials by contact spotting or spray-on technique. The ADC provides proper timing of chamber saturation, conditioning, development, and drying of the TLC plate. It also adjusts the relative humidity which provides reproducible results. HPTLC visualization provides photo documentation of TLC fingerprints in the form of fingerprint images. TLC scanner provides scanning from 190-800 nm that can be used for the identification of compounds.

CHAPTER 3

Process Steps for HPTLC Analysis

Abstract: This chapter briefly discusses the general aspects that can affect the result of TLC. Sample preparation, application of samples, preparation of development solvents, the steps involved in the development of TLC, visualization and derivatization, and detection of analyte are the key steps of chromatographic development. This chapter focuses on the process steps for HPTLC analysis.

Keywords: Derivatization, Detection, Stationary phase, Visualization.

INTRODUCTION

Sample and standards

Appropriate sample preparation is an essential step for successful TLC analysis. A suitable solvent system was used for either dissolving or extracting a solid or liquid sample to make the target molecules present in the sample available for analytical procedures without any interference from the sample matrix.

Preparation of Sample Solution

The raw materials or extracts of the medicinal herbs were processed and extracted using the appropriate solvent system, *i.e.* water/ethanol/methanol/hydro-methanol/ethyl acetate/hexane. Approximately, 1000 mg (for raw material) or 500 mg (for extract) of the test sample was transferred to a 10 mL volumetric flask and 5 mL of the solvent was added and homogenized in an ultrasonic bath for 20 minutes. The volume of the solution was made up to 10 mL mark with the solvent. The samples were centrifuged for 10 minutes at 9000 RPM and the supernatant, devoid of any particles was transferred to the sample application tube.

Standard Solution Preparation

For the preparation of the standard solution, 10 mg of the commercially available standard was weighed and transferred into a 10 mL volumetric flask. The standard

Acharya Balkrishna

was dissolved in 5 mL methanol, sonicated for 5 minutes and the volume of the standard was made up to 10 mL with methanol to obtain a concentration of 1 mg/mL.

Stationary Phase

The stationary phase is an adsorbent, material layer, usually silica gel or aluminum oxide, coated onto a surface, which is typically a glass plate or aluminum. There are a number of inert materials that are available as sorbent in TLC. To achieve optimum separation, it is very important to choose the correct material. Silica gel and aluminum oxide is mostly used for a wide range of applications. Silica gel and aluminas can also be split into a number of distinct sorbents, depending on the pore size, particle size and pH. The selection of the sorbent depends upon the nature and the type of compounds to be separated. Polarity, solubility, and molecular weight are some of the characteristic features for deciding the separation mechanism [7].

TLC silica gel 60 F_{254} (Merck KGaA, 64271 Darmstadt Germany) aluminum backed plates with a dimension of 5 × 10 cm or 10 × 10 cm or 20 × 10 cm (width × length) were used as stationary phase [7].

Sample Application

Application of sample and/or reference solution is a very critical step for obtaining quality results. This is due to the fact that precise and uniform sample application determine the overall quality of resolution and eventually the quantification values.

Band-wise spray-on technique was used for sample application on TLC plates. Suitable volumes of sample and standard were applied as 8 mm narrow bands on TLC plates using CAMAG ATS4, equipped with a 25 µL Hamilton syringe programmed by winCATS software.

Preparation of Developing Solvent

The mobile phase or developing solvent is the liquid that flows through a chromatographic system, moving the material to be separated at different rates on the stationary phase [6].

Development solvents consisting of more than one component were prepared by measuring the required volume of each component separately and transferring them into a solvent bottle of appropriate size. The bottle is closed with a lid and shaken to ensure proper mixing of the content. For measuring volumes smaller than 1 mL, a suitable micropipette was used, and a graduated volumetric pipette of suitable size was used for volumes up to 20 mL. A list of commonly used solvent systems is provided in Appendix Table A1.

Development of TLC Plate

TLC plates were developed in a saturated twin trough chamber (TTC) according to the following procedure:

1. Appropriate volume of developing solvent was prepared (10 mL for 10 × 10 cm, 20 mL for 20 × 10 cm TTC).
2. A piece of correct sized (10 × 10 cm, 20 × 10 cm) filter paper was placed inside the rare trough.
3. The solvent was transferred into a chamber so that the filter paper is thoroughly wetted and adheres to the rare wall of TTC.
4. Chamber was tilted to the side (about 45 degrees) to equalize the volume of solvent in both troughs.
5. The chamber was allowed to saturate for 10 to 15 minutes or more on the basis of method development.
6. The TLC plate was then placed inside the TTC and allowed to develop up to 70 mm followed by drying the developed TLC plate.

Visualization and Derivatization

After development, compounds can be visualized under white light, long UV light (366 nm), short wave UV light (254 nm), or in white light after derivatization Fig. (**11**). Derivatization is performed to make compounds visual that cannot be visualized under UV light. There are several derivatizing reagents, which can be used for the purpose based upon the nature of compounds [6]. A list of commonly used derivatizing reagents is provided in Appendix Table A2.

For derivatization, the developed TLC plates were sprayed with derivatizing reagents and heated using a TLC heater at 110°C for 5-10 minutes.

Detection

Detection can be performed in absorbance or fluorescent mode using a tungsten, deuterium or mercury lamp. Deuterium lamp offers a continuous spectrum of UV light from 190 to 400 nm, tungsten lamp at 370 to 800 nm, and mercury vapour lamp at 366 nm. Detection generates chromatograms that can be used for both qualitative and quantitative purposes Fig. (**12**).

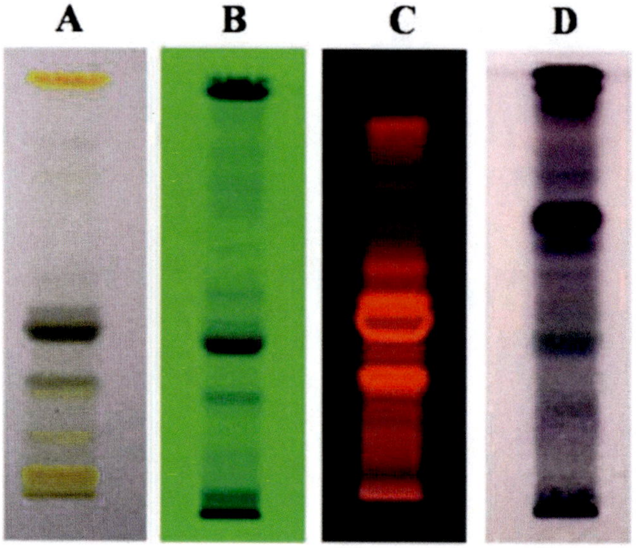

Fig. (11). HPTLC fingerprint of *Cinnamomum camphora* (L.) J.Presl. Mobile phase (MP): Hexane: ethyl acetate (8: 2 v/v). A: Under white light before derivatization, B: 254 nm, C: 366 nm, D: Under white light after derivatization.

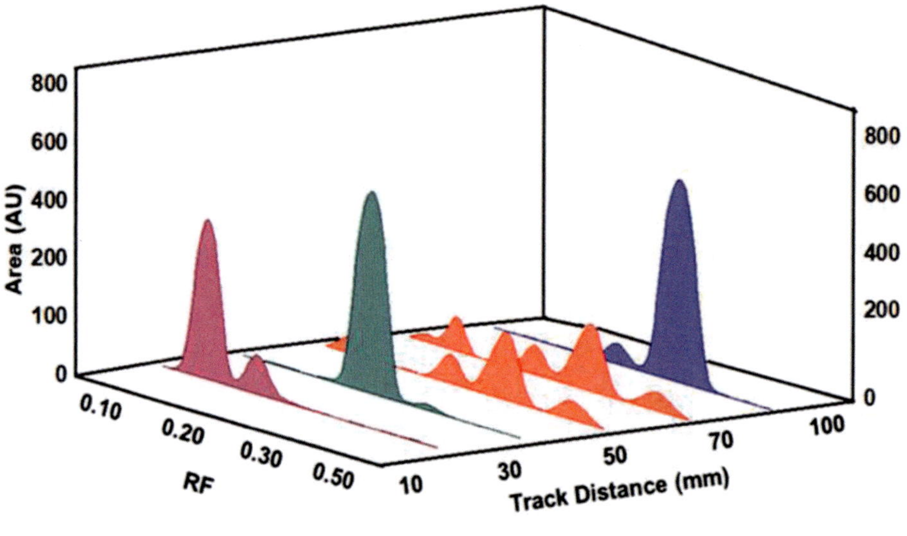

Fig. (12). Three-dimensional plot of sample scanned at 280 nm using deuterium lamp in absorbance reflection mood.

CONCLUDING REMARKS

HPTLC provides the opportunity to visualize the results directly in the form of bands on the TLC plate. The bands of compounds can be easily visual at short UV, and long UV under white light with or without derivatization, on the basis of the UV absorbance of the compound. The results can be presented in the form of a TLC fingerprint. We can also do quantitative analysis in HPTLC by using a scanner at the required wavelength. HPTLC provides both qualitative and quantitative analysis facilities.

CHAPTER 4

Stationary Phase in High Performance Thin Layer Chromatography (HPTLC)

Abstract: As discussed in the first chapter, Chromatography is a separation technique based on the different affinities of the components in a mixture of two immiscible phases. One phase does not move (stationary phase) and the other moves through it in a definite direction (mobile phase). This chapter discusses the adsorbent or stationary phase used in TLC and the properties of these materials.

Keywords: Aluminium, Adsorption, Stationary phase, Silica gel.

INTRODUCTION

In TLC, the stationary phase is a thin layer of adsorbing materials, mostly silica gel or aluminum oxide, coated upon an inert glass, plastic, or aluminium plate surface. From the point of view of chemicals, silica gel is silicium dioxide, where each silicium atom is bound with four atoms of oxygen [4].

There are a number of materials that can be used as adsorbent in TLC. Some adsorbing materials can be used for all types of compounds (silica gel), while some have a limited range of applications (silica gel impregnated with caffeine for polyaromatic hydrocarbons). When we select the sorbent for analysis, the focus should be given to the class of compounds to be analyzed or separated. The polarity, solubility, molecule weight, shape and size of the targeted molecule play an important role in deciding the mechanism of separation. The separation of components depends on both the adsorbing material and solvent system [5].

The most widely used sorbent is silica gel, followed by cellulose and alumina. With time, there is an evolution in silica gel also, which results in opening up new separation. Some sorbents that can be used on the basis of class of compounds are shown below (4) (Table **1**).

SILICA GEL

Silica gel is also known as silicic acid and kieselgel gel. From the chemical point, silica gel is silicium dioxide, where each silicium atom is bound with four atoms

Acharya Balkrishna

of oxygen. The process of formation of silica gel starts with sand. When sand reacts with sodium hydroxide or sodium carbonate, it is converted into soluble sodium silicate. When this silicate solution is acidified, it gets hydrolyzed to form silicic acid which condensates to very tiny polysilicic acid hydrosol. With time, the hydrogel gets aggregated, establishes siloxane bonds and takes the shape of a cake. This cake is pulverized and washed to remove soluble impurities and salts of sodium [5].

Table 1. List of some commonly used adsorbing materials used as stationary phases.

S. NO.	Adsorbing Material	Class of Compounds
1	Silica gel	All types of compounds: Phenols, alkaloids, carbohydrates, aromatic compounds.
2	Aluminium oxide	Alkaloids, amines, steroids, aromatic compounds, aliphatic hydrocarbons.
3	Cellulose	Amino acids, carbohydrates
4	Reversed phase silica gel	Fatty acids, lipids, tetracycline's

The size and distribution of silica gel are primary properties of TLC. Smaller particles with narrow size distribution yield highly efficient chromatographic beds. Silica gel used for TLC usually has a particle size of 15 to 20 μm. HPTLC plates are prepared with smaller particles of silica gel ranging from 5 to 7 μm.

Pore size plays an important role in the chromatographic performance of silica gel. TLC material has an average pore size of 60 Å or 6 nm that is why the term Si 60.

Silica is porous, hydrated silicon dioxide, where each silicon atom is surrounded with four oxygen atoms. Some of these bonds are present on the surface of the pores [4]. These structures under chromatographic conditions allow silica gel to interact with analytes of different types. Some following types of interactions are possible:

- Proton donor or acceptor interactions.
- Dipole dipole interactions.
- Interaction with dipoles induced by the stationary phase.
- Interactions based on dispersion forces.

Silica Gel Bonded Phases

Silica gel provides a wide range of possibilities to separate different classes of compounds on silica gel. However, there are some bonded phases that can be grouped into polar and non-polar phases (Table **2**).

Some non-polar and polar bonded phases:

Table 2. List of some polar and non-polar bonded phases.

S. No.	Name	Functionality	Polarity
1	RP_2	Dimethyl	Non polar
2	RP_8	Octyl	Non polar
3	RP_{18}	Octadecyl	Non polar
4	Whatman	Diphenyl	Non polar
5	Amino	3- Aminopropyl	Polar
6	Cyano	3- Cyanopropyl	Polar
7	Diol	Spacer bonded propanediol	Polar

These bonded phases can be produced easily. By boiling silica gel in water, number of free hydroxyl groups can be generated. The silica gel is than hydrolyzed using inert organic solvent for example C18 chain. On the basis of the type of silane, the bonding, the surface coverage and degree of cross linking also vary.

Some other commonly used stationary phase sorbents are aluminium oxide and cellulose. Aluminium oxide is prepared by dehydration of aluminium hydroxide at a high temperature. It contains acidic aluminium ions, basic oxide ions and polar AI-OH group. This is useful for the separation of alkaloids, amines, steroids, aromatic compounds, and aliphatic hydrocarbons.

Cellulose is available in microcrystalline or in fibrous form. As it is a carbohydrate, it works well with water and lower alcohols. It provides good separation when the solvent system is aqueous or water based. It can also separate very polar compounds, but its migration speed is very slow. It takes a lot of time to reach the solvent front. It is used for amino acids, and carbohydrates separation.

CONCLUDING REMARKS

The stationary phase plays an important role in the separation of compounds, therefore, the selection of the stationary phase on the basis of the analyte is very important. Silica gel is the most used and versatile stationary phase that can be used to separate different classes of compounds. Silica gel is also available which can be used for specific targeted compounds. Other than silica, aluminium oxide and cellulose are also used as stationary phases.

CHAPTER 5

Hyphenation of HPTLC with other Techniques

Abstract: Hyphenation techniques are those where two different analytical techniques are combined together to give synergy. The advantage includes high separation efficiency, and the detection of compounds with high sensitivity and specificity. HPTLC can be combined with liquid chromatography (LC), infrared (IR) spectroscopy, ultra violate (UV) spectroscopy, and mass spectroscopy (MS). This chapter briefly discusses the hyphenation of HPTLC with other analytical techniques, its uses, advantages and areas of application of HPTLC coupled with other analytical techniques.

Keywords: HPTLC, LC, IR, MS, MS-Interface, Hyphenation.

INTRODUCTION

A hyphenation is a combinational approach, which means combining one technique with another. The combination of two different techniques can lead to the production of a large number of compounds with a wide range of spectral diversity in a short span of time. HPTLC can be hyphenated with high-pressure liquid chromatography (HPLC), mass spectroscopy (MS), Fourier transform infrared spectroscopy (FTIR) and Raman spectroscopy.

High-performance Thin Layer Chromatography (HPTLC) Hyphenated with Mass Spectroscopy (MS)

HPTLC in combination with MS, brings an added level of information to the chromatographic results and it also makes the identification and detection of molecules much easier. HPTLC coupled with MS allows verification of the chemical structure of molecules. By using the HPTLC-MS-Interface Fig. (**13**), the analyte can be directly eluted out from the developed TLC plate.

Compounds are separated on the TLC plate, after that the targeted molecules are marked either under UV 254 nm, 366 nm or white light. TLC is manually positioned on HPTLC-MS, the piston is then lowered by pressing the push button. The piston should lock the marked or targeted zone. Pushing the lever starts the solvent flow.

Acharya Balkrishna

The solvent passes through the locked zone on TLC, and then the molecule is dissolved with the solvent that can directly be injected into MS or collected in a separate sample storage tube. Hyphenating HPTLC with MS provides new dimensions for the technique and makes it more prestigious from a scientific point of view. HPTLC-MS can be used for screening phyto-metabolites, identification of bio-active molecules, and detection and identification of unknown impurities present in samples [4].

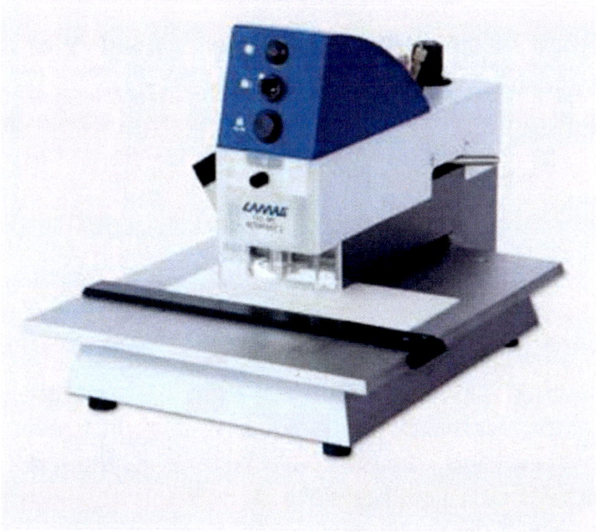

Fig. (13). CAMAG HPTLC-MS-Interface.

In the field of medicinal plants, this technique can be very useful, as the herbal plant material and extracts contain a very complex matrix, there are a number of compounds present in herbal materials, and isolation of these compounds by column chromatography takes a lot of time. Moreover, the isolation of each compound is not possible. By using TLC, the compounds can be separated and separated molecules can be easily identified by HPTLC-MS Fig. (**14**).

High Performance Thin Layer Chromatography (HPTLC) Hyphenated with Infrared Spectroscopy (IR)

HPTLC in combination with IR can be a great technique for the identification of the molecule or the functional group present in the sample. IR works on the principle that when IR radiation passes through the sample, some of the radiation is absorbed, and the radiation that passes by the sample is recorded in the form of spectra. As different molecule has different spectra, and on the basis of spectra, the molecule can be easily identified.

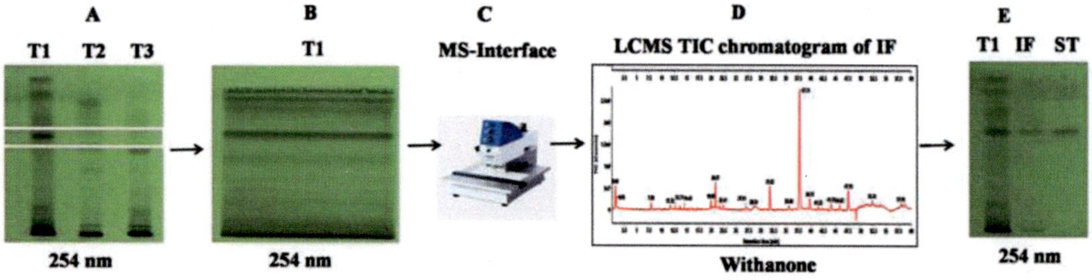

Fig. (14). Showing the application of HPTLC-MS-Interface in the field of medicinal plants for the identification of compounds. HPTLC fingerprint of *Withania somnifera* (Ashwagandha) at 254 nm along with the LCMS TIC chromatogram of isolated compound (by MS-Interface) and TLC of that isolated compound with withanone standard. M.P.: Ethyl acetate: toluene: formic acid (10: 1: 1 v/v/v). A: TLC fingerprint of T1: Ashwagandha mature leaves extract, T2: Ashwagandha juvenile leaves extract, T3: Ashwagandha whole plant extract. The marked compound is only present in T1 (Ashwagandha mature leaves extract), B: Preparative TLC of T1 (Ashwagandha mature leaves extract), C: MS-Interface for collection of the marked compound from TLC, D: LCMS TIC chromatogram of the collected compound (which is identified as withanone on the basis of its mass), E: HPTLC fingerprint of (T1) Ashwagandha mature leaves extract, (T2) Isolated fraction from T1 by MS-Interface, (ST) Standard withanone which is re-confirming the identified compound is withanone.

The HPTLC IR coupling can be categorized into two groups, direct and indirect hyphenation. Direct method is based on the direct hyphenation of HPTLC-FTIR, while in the indirect method, the targeted zone is collected from the TLC and transferred to IR. The spectra is then recorded. The HPTLC-FTIR, spectra make possible the detection and quantification of UV as well as non-UV compounds on the TLC plate.

Other analytical techniques that can be hyphenated with HPTLC are LC, UV spectroscopy and laser. Coupling of this technique can make the detection of molecules more simple and advanced. Coupling of HPTLC with other analytical techniques can be useful in food, chemical, biology, environmental and herbal or natural product areas.

CONCLUDING REMARKS

HPTLC is one of the most versatile techniques. It can easily handle various types of samples without consuming a lot of time. The hyphenation of HPTLC with other analytical techniques such as, UV, FTIR, LC and MS open various ways of the detection of molecules. It also gives new dimensions and makes hyphenation more prestigious from scientific point of view.

CHAPTER 6

Fingerprints of Medicinal Plants

Abstract: Medicinal plants are considered a rich source of phytochemicals, which can be used in drug development. For this purpose, the selection of correctly identified herbs is of significance as each herb has a unique phytochemical distribution pattern. High-Performance Thin Layer Chromatography (HPTLC) is one of the simplest yet robust techniques for the identification of herbs, as it can provide unique phytochemical fingerprints for each herb. However, process parameters of generating HPTLC fingerprints require rigorous optimization for each plant. This chapter provides HPTLC fingerprints for 108 medicinal plants together with their botanical name and common Hindi names. Besides providing optimized HPTLC conditions for generating these unique 108 fingerprints, this book also offers a list of solvent systems and derivatizing reagents for screening of phytochemicals. With classical taxonomical features and HPTLC fingerprinting methodology compiled, this chapter aims to serve as a guide for the identity verification of medicinal plants.

Keywords: Adsorbent, Canvas painting, HPTLC fingerprint, Medicinal Value, Solvent system.

INTRODUCTION

HPTLC fingerprint is a chromatographic pattern of phyto-active molecules that is specific for each and every medicinal plant. Chemical fingerprints provide an accurate method for the identification and authentication of herbal plant materials.

Achyranthes aspera L. (Common Name: Apamarga)

Plant Part: Whole plant.

Solvent System: Chloroform: methanol (9: 1 v/v).

Adsorbent: TLC silica gel 60 F_{254} (Merck).

Standard Solution Preparation: 10 mg of oleanolic acid was dissolved in 10 mL of methanol, and 10µL of the standard solution was applied on the TLC plate.

Acharya Balkrishna

Sample Solution Preparation: 100 mg of sample was dissolved in 10 mL methanol, sonicated and centrifuged, and 20 µL of the solution was applied on the TLC plate.

Derivating Reagent: Anisaldehyde sulfuric acid.

Medicinal Value: Useful in urinary ract disorders, indigestion and cold (1).

Conclusion: Rf 0.51 confirms the presence of bio-active marker oleanolic acid in Apamarga by HPTLC fingerprinting (Fig. **15a**, **b**).

Fig. (15a). Canvas painting of Apamarga illustrated by Patanjali Research Foundation, Haridwar, India.

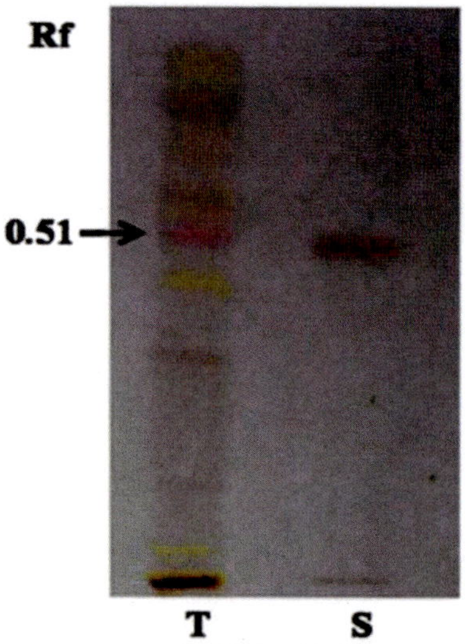

Fig. (15b). HPTLC fingerprinting of Apamarga: T: Apamarga whole plant, S: Oleanolic acid.

Acorus calamus L. (Commmon Name: Vach)

Plant Part: Rhizome.

Solvent System: Toluene: ethyl acetate: formic acid (5: 5: 1 v/v/v).

Adsorbent: TLC Silica gel $60F_{254}$ (Merck).

Sample Solution Preparation: 1000 mg sample dissolved in 10 mL of methanol, sonicated and centrifuged. 10µl of solution was applied on the TLC plate.

Derivatizing Reagent: Anisaldehyde sulfuric acid.

Medicinal Value: Used in skin diseases, respiratory and gastrointestinal diseases (3).

Conclusion: HPTLC fingerprint of Vach shows bands of phytochemicals under white light after derivatization (Fig. **16a**, **b**).

Fig. (16a). Canvas painting of Vach illustrated by Patanjali Research Foundation, Haridwar, India.

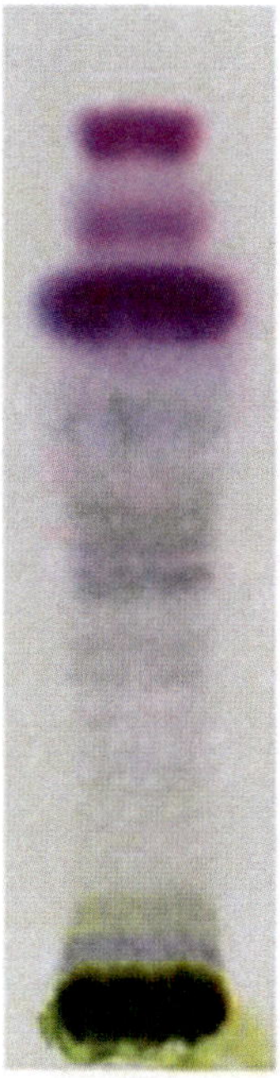

Fig. (16b). HPTLC fingerprinting of Vach: T: Vach rhizome.

Aegle marmelos (L.) Correa (Commmon Name: Bael)

Plant Part: Fruit.

Solvent System: Toluene: ethyl acetate: formic acid (10: 10: 1 v/v/v).

Adsorbent: TLC silica gel 60 F_{254} (Merck).

Sample Solution Preparation: 100 mg of sample dissolved in 10 mL methanol, sonicated and centrifuged. 20 μL of solution was applied on TLC plate.

Medicinal Value: The plant has antimicrobial, antipyretic, ulcer healing, diuretic and anti-inflammatory properties (2).

Conclusion: Fingerprinting of Bael showing specific pattern of bands at 254 nm and 366 nm (Fig. **17a**, **b**).

Fig. (17a). Canvas painting of Bael illustrated by Patanjali Research Foundation, Haridwar, India.

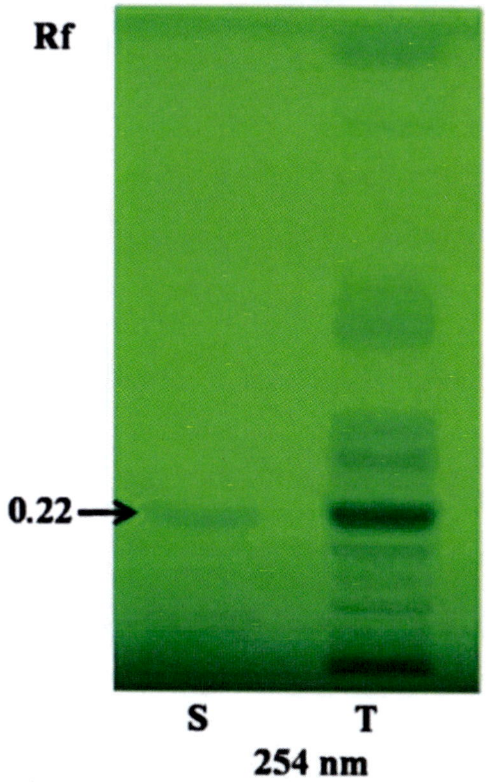

Fig. (17b). HPTLC fingerprinting of Bael: T: Bael fruit.

Albizia odoratissima (L.f.) Benth (Common Name: Siris)

Plant Part: Bark.

Solvent System: Toluene: ethyl acetate: formic acid (10: 10: 1 v/v/v).

Adsorbent: TLC silica gel 60 F_{254} (Merck).

Sample Solution Preparation: 100 mg of sample dissolved in 10 mL methanol, sonicated and centrifuged, and 15 µL of the solution was applied on a TLC plate.

Medicinal Value: Useful in skin disorders, cough, cold and blood disorders (3).

Conclusion: Fingerprinting of Siris shows a major fluorescent band and numbers of bands at 366 nm (Fig. **18a**, **b**).

Fig. (18a). Canvas painting of Siris illustrated by Patanjali Research Foundation, Haridwar, India.

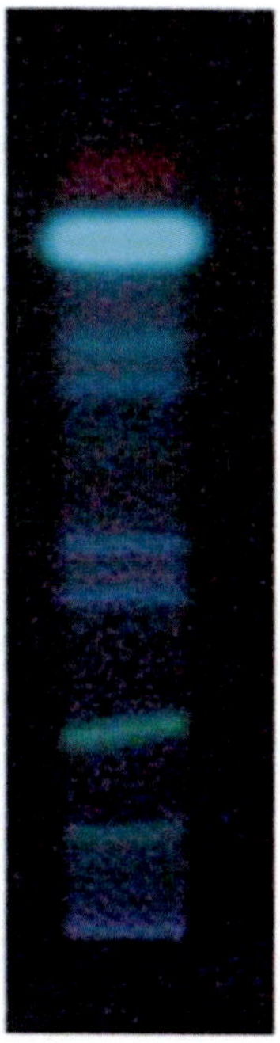

Fig. (18b). HPTLC fingerprinting of Siris: T: Siris bark.

Allium sativum L. (Common Name: Garlic)

Plant Part: Bulb.

Solvent System: Toluene: ethyl acetate: acetic acid: formic acid (5: 3: 1: 0.5 v/v/v/v).

Adsorbent: TLC Silica gel $60F_{254}$ (Merck).

Sample Solution Preparation: 1000 mg sample dissolved in 10 mL methanol, sonicated and centrifuged. 10 µL of solution was applied on a TLC plate.

Medicinal Value: Improves digestion, boosts immunity, and improves skin health (3).

Conclusion: HPTLC fingerprint of Garlic shows spots of phyto-chemicals under 254 nm and 366 nm (Fig. **19a**, **b**).

Fig. (19a). Canvas painting of Garlic illustrated by Patanjali Research Foundation, Haridwar, India.

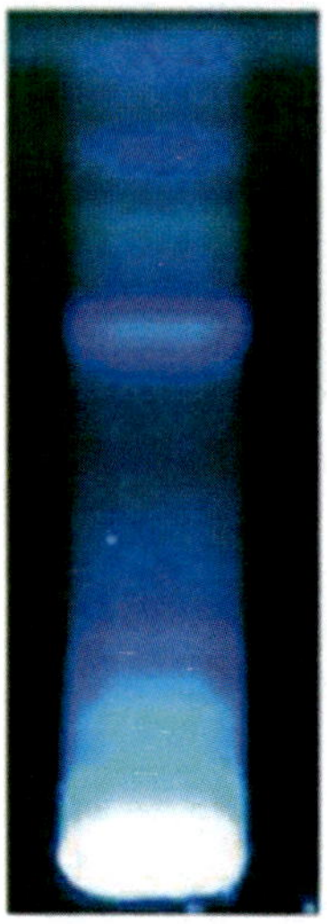

T

366 nm

Fig. (19b). HPTLC fingerprinting of Garlic: T: Garlic bulb.

Aloe barbadensis Mill (Common Name: Aloe Vera)

Plant Part: Leaf gel.

Solvent System: Ethyl acetate: methanol: water (10: 1.35: 1 v/v/v).

Adsorbent: TLC Silica gel $60F_{254}$ (Merck).

Standard Solution Preparation: 10 mg of aloin was dissolved in 10 mL of methanol. 10 μL of standard solution was applied on a TLC plate.

Sample Solution Preparation: 1000 mg of the sample was dissolved in 1 mL water and fractionated with 1 mL ethyl acetate. Ethyl acetate layer is collected and dried and further re-constituted in 1 mL methanol. 6 μL of sample solution was applied on TLC plate.

Medicinal Value: Useful in treating acne and burns, and improves digestion and oral health (1).

Conclusion: HPTLC fingerprinting of Aloe Vera represents the presence of signature molecule aloin at Rf 0.60, with respect to the standard aloin (Fig. **20a**, **b**).

Fig. (20a). Canvas painting of Aloe Vera illustrated by Patanjali Research Foundation, Haridwar, India.

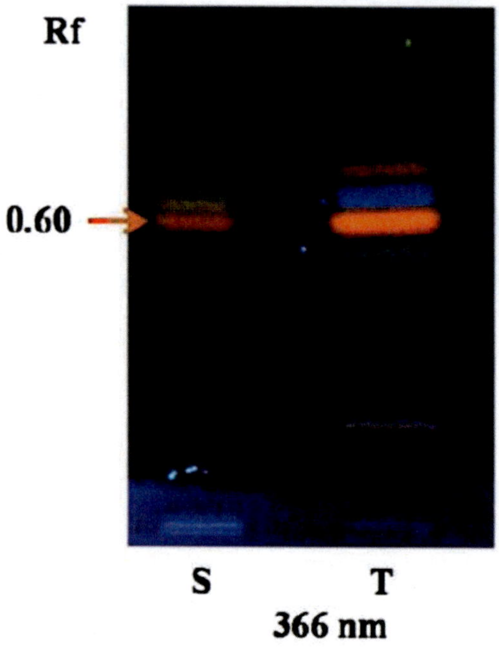

Fig. (20b). HPTLC fingerprinting of Aloe vera: S: Aloin, T: Aloe vera gel.

***Anacyclus pyrethrum* (L.) Lag (Common Name: Akarkara)**

Plant Part: Roots.

Solvent System: Toluene: ethyl acetate: formic acid (5: 5: 1 v/v/v).

Adsorbent: TLC Silica gel $60F_{254}$ (Merck).

Sample Solution Preparation: 1000 mg sample dissolved in 10 mL of methanol, sonicated and centrifuged. 2 μL of solution was applied on TLC plate.

Medicinal Value: Used to treat asthma, throat pain, stomach ache, and tooth and gum problems (1).

Conclusion: HPTLC fingerprint of Akarkara shows bands of phytochemicals under 254 nm and 366 nm (Fig. **21a**, **b**).

Fig. (21a). Canvas painting of Akarkara illustrated by Patanjali Research Foundation, Haridwar, India.

Fig. (21b). HPTLC fingerprinting of Aakarkara: T: Akarkara roots.

Andrographis paniculata (Burm. F.) Nees (Common Name: Kalmegh)

Plant Part: Leaves.

Solvent System: Toluene: ethyl acetate: formic acid (6: 4: 0.3v/v/v).

Adsorbent: TLC Silica gel $60F_{254}$ (Merck).

Sample Solution Preparation: 1000 mg of sample dissolved in 10 mL of ethanol, sonicated and centrifuged. 20 μL of solution was applied on a TLC plate.

Derivatizing Reagent: Anisaldehyde sulfuric acid.

Medicinal Value: Useful in indigestion, fever, cough, cold and diarrhoea (1).

Conclusion: HPTLC fingerprint of Kalmegh captured at 254 nm, 366 nm and white light, showing bands of phytochemical compounds under UV and visible region (Fig. **22a**, **b**).

Fig. (22a). Canvas painting of Kalmegh illustrated by Patanjali Research Foundation, Haridwar, India.

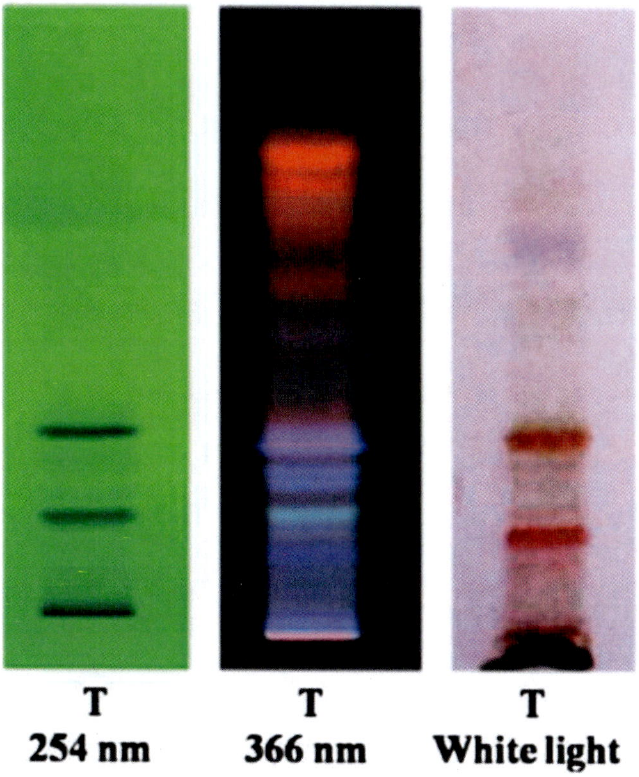

Fig. (22b). HPTLC fingerprinting of Kalmegh: T: Kalmegh leaves.

Areca catechu L. (Common Name: Betel Nut)

Plant Part: Fruit.

Solvent System: Toluene: ethyl acetate: formic acid (7: 3: 1 v/v/v).

Adsorbent: TLC Silica gel $60F_{254}$ (Merck).

Sample Solution Preparation: 1000 mg of sample dissolved in 10 mL of methanol, sonicated and centrifuged. 2 μL of solution was applied on a TLC plate.

Derivatizing Reagent: Anisaldehyde sulfuric acid.

Medicinal Value: Used as mouth freshener, improves salivation, and digestion. Also have anti-fatigue, antibacterial and antifungal properties.

Conclusion: HPTLC fingerprint of Betel nut shows intense purple colored bands in white light after derivatization (Fig. **23a**, **b**).

Fig. (23a). Canvas painting of Betel nut illustrated by Patanjali Research Foundation, Haridwar, India.

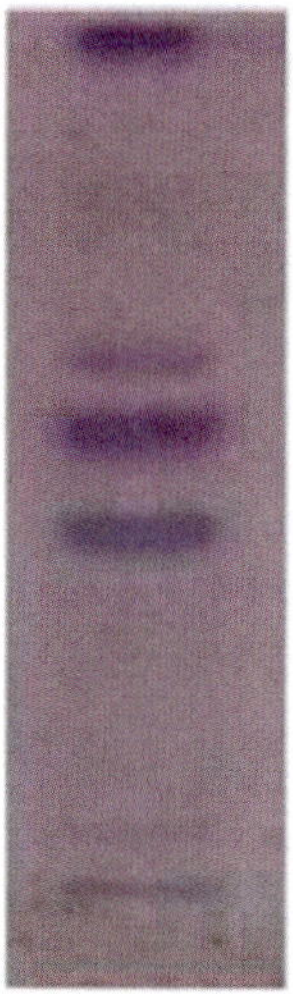

T
White light

Fig. (23b). HPTLC fingerprint of Betel nut: T: Betel nut fruit.

Asparagus racemosus Willd (Common Name: Shatavar)

Plant Part: Root.

Solvent System: Toluene: ethyl acetate: formic acid: methanol (8: 8: 2: 2 v/v/v/v).

Adsorbent: TLC Silica gel $60F_{254}$ (Merck).

Sample Solution Preparation: 1000 mg of sample dissolved in 10 mL of methanol, sonicated and centrifuged. 8µL of solution was applied on a TLC plate.

Derivatizing Reagent: Anisaldehyde sulfuric acid

Medicinal Value: Used to treat cough, diarrhoea, and ulcer. It is diuretic and boosts immunity (3).

Conclusion: HPTLC fingerprint of Shatavar shows major fluorescent band at 366 nm and a specific pattern of bands in white light after derivatization (Fig. **24a**, **b**).

Fig. (24a). Canvas painting of Shatavar illustrated by Patanjali Research Foundation, Haridwar, India.

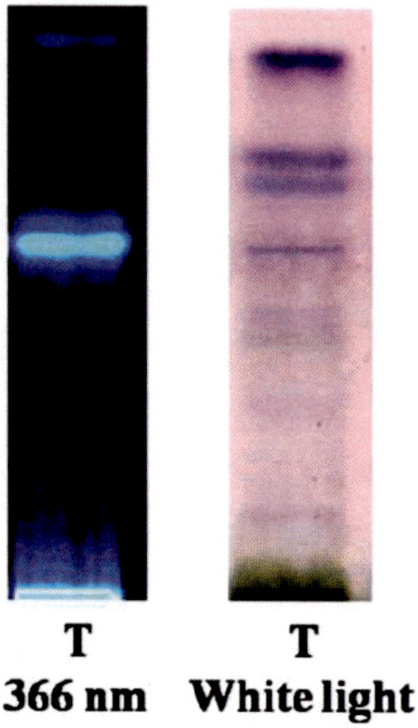

Fig. (24b). HPTLC fingerprinting of Shatavar: T: Shatavar roots.

Azadirachta indica A.Juss (Common Name: Neem)

Plant Part: Leaves.

Solvent System: Toluene: ethyl acetate: formic acid (5: 5: 1 v/v/v).

Adsorbent: TLC Silica gel $60F_{254}$ (Merck).

Sample Solution Preparation: 1000 mg of sample dissolved in 10 mL of methanol, sonicated and centrifuged. 10µL of solution was applied on a TLC plate.

Medicinal Value: Used to treat acne, fungal and bacterial infections, and prevent gastrointestinal diseases (2).

Conclusion: HPTLC fingerprint of Neem represents a pattern of fluorescent bands at 366 nm and other bands of phytochemicals at 254 nm (Fig. **25a**, **b**).

Fig. (25a). Canvas painting of Neem illustrated by Patanjali Research Foundation, Haridwar, India.

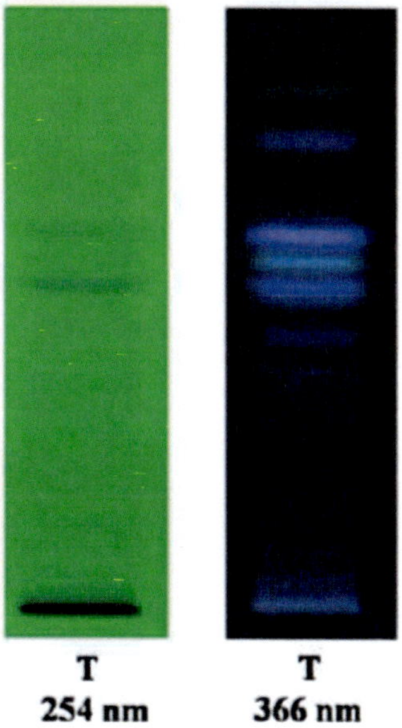

Fig. (25b). HPTLC fingerprint of Neem: T: Neem leaves.

Baccharoides anthelmintica **(L.) Moench (Common Name: Black Cumin)**

Plant Part: Seeds.

Solvent System: Toluene: Ethyl acetate: Formic acid (6: 5: 1 v/v/v).

Adsorbent: TLC Silica gel $60F_{254}$ (Merck).

Sample Solution Preparation: 500 mg sample dissolved in 10 mL ethanol, sonicated and centrifuged. 12 μL of solution was applied on a TLC plate.

Medicinal Value: Used to treat threadworms, indigestion, cough and cold.

Conclusion: HPTLC fingerprint of Black cumin shows numbers of bands at 254 nm and 366 nm. This fingerprint is specific for the sample in given chromatographic conditions (Fig. **26a**, **b**).

Fig. (26a). Canvas painting of Black cumin illustrated by Patanjali Research Foundation, Haridwar, India.

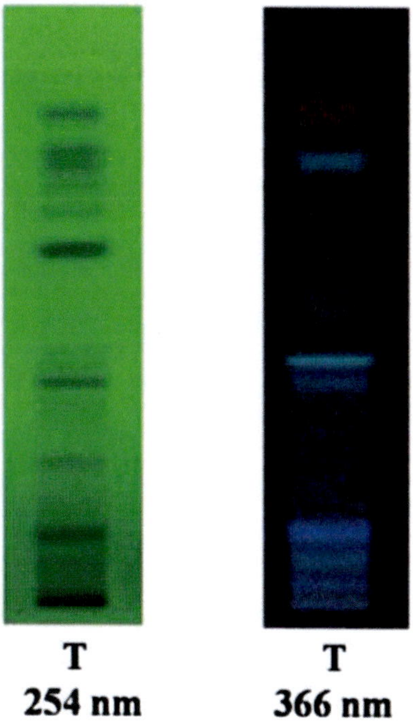

Fig. (26b). HPTLC fingerprinting of Black cumin: T: Black cumin seeds.

***Bacopa monnieri* (L.) Wettst (Common Name: Brahmi)**

Plant Part: Aerial part.

Solvent System: Hexane: ethyl acetate (8: 2 v/v).

Adsorbent: TLC Silica gel $60F_{254}$ (Merck).

Sample Solution Preparation:1000 mg sample dissolved in 10 mL of methanol, sonicated and centrifuged. 10µL of solution was applied on a TLC plate.

Derivatizing Reagent: Anisaldehyde sulfuric acid.

Medicinal Value: Used to boost brain functions, and reduce anxiety. It contains antioxidants (2).

Conclusion: HPTLC fingerprint, displays very prominent red colored bands in 366 nm and bands are also visible at 254 nm and under white light after derivatization (Fig. **27a**, **b**).

Fig. (27a). Canvas painting of Brahmi illustrated by Patanjali Research Foundation, Haridwar, India.

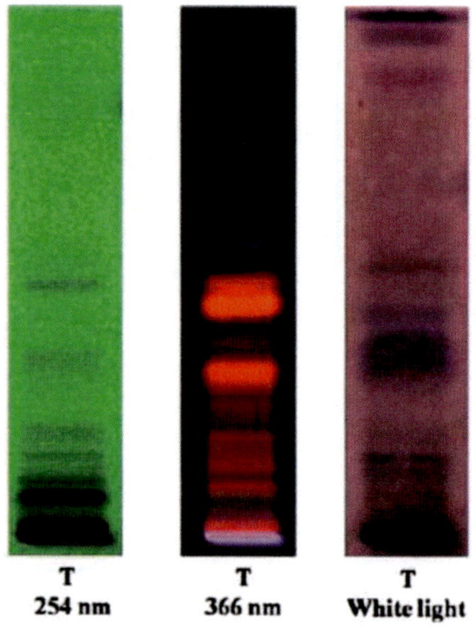

Fig. (27b). HPTLC fingerprinting of Brahmi: T: Brahmi aerial part.

Bauhinia variegata **L. (Common Name: Kachnar)**

Plant Part: Bark.

Solvent System: Toluene: ethyl acetate: formic acid (5: 2: 1 v/v/v).

Adsorbent: TLC Silica gel $60F_{254}$ (Merck).

Sample Solution Preparation: 1000 mg sample dissolved in 10 mL of methanol, sonicated and centrifuged. 10 µL of solution was applied on a TLC plate.

Derivatizing Reagent: Anisaldehyde sulfuric acid.

Medicinal Value: It promotes wound healing, reduce inflammation, cure cough and indigestion (1).

Conclusion: HPTLC fingerprint of Kachnar shows specific banding pattern at 254 nm and 366 nm and under white light after derivatization (Fig. **28a**, **b**).

Fig. (28a). Canvas painting of Kachnar illustrated by Patanjali Research Foundation, Haridwar, India.

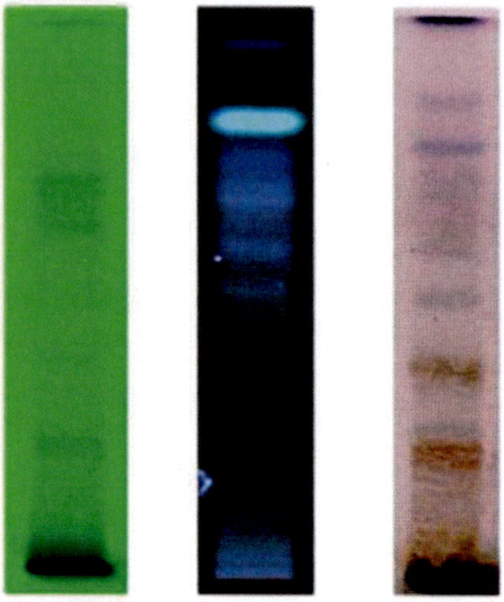

Fig. (28b). HPTLC fingerprint of Kachnar: T: Kachnar bark.

Berberis aristata DC (Common Name: Daruhaldi)

T2: Daruhaldi stem water extract

Plant Part: Stem.

Solvent System: Butanol: acetic acid: water (6.5: 1.5: 2 v/v/v).

Adsorbent: TLC Silica gel $60F_{254}$ (Merck).

Standard Solution Preparation: 1 mg of berberine was dissolved in 10 mL of methanol. 10 μL of standard solution was applied on a TLC plate.

Sample Solution Preparation: 500 mg of each sample dissolved in 10 mL of respective solvents, sonicated and centrifuged. 15 μL of solution was applied on TLC plate.

Medicinal Value: Used in treating fever, wound healing, inflammation, cold, skin diseases (2).

Conclusion: HPTLC fingerprint validated the presence of bio-active marker berberine in Daruhaldi. Samples showing the band of berberine at 0.45 Rf, similar to the berberine standard at 254 nm and 366 nm (Fig. **29a**, **b**).

Fig. (29a). Canvas painting of Daruhaldi illustrated by Patanjali Research Foundation, Haridwar, India.

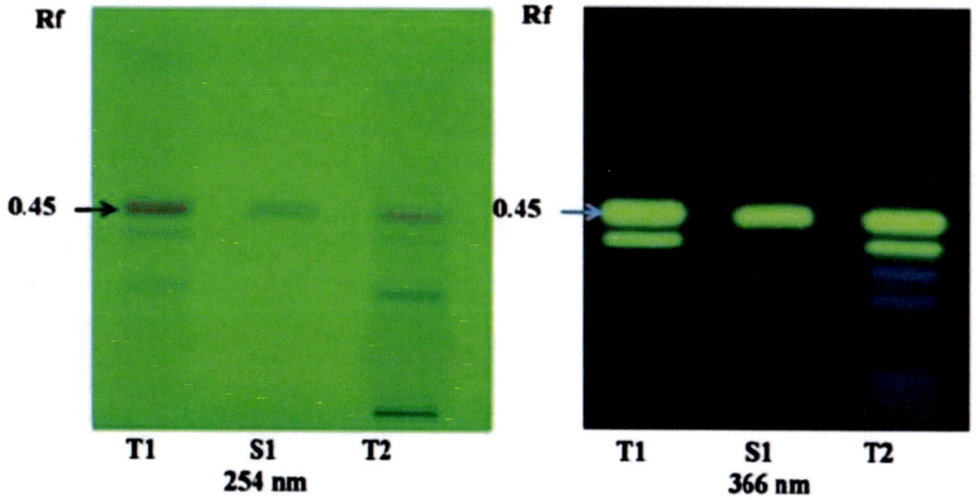

Fig. (29b). HPTLC fingerprint of Daruhaldi: S1: Berberine, T1: Daruhaldi stem alcoholic extract,.

Berberis napaulensis (DC.) Spreng (Common Name: Mahonia)

Plant Part: Stem.

Solvent System: Butanol: acetic acid: water (6.5: 1.5: 2 v/v/v).

Adsorbent: TLC Silica gel $60F_{254}$ (Merck).

Standard Solution Preparation: 10 mg of each berberine and palmatine were dissolved in 10 mL of methanol. 1µL of standard solution was applied on a TLC plate.

Sample Solution Preparation: 500 mg of the sample dissolved in 10 mL of methanol, sonicated and centrifuged. 1µL of solution was applied on a TLC plate.

Medicinal Value: Useful in treating fever, wound healing, inflammation, cold, and skin disease.

Conclusion: HPTLC fingerprint validates the presence of signature markers in Mahonia sample, showing the bands of berberin and palmatine at 0.45 and 0.40 Rf respectively, similar to the standard at 366 nm (Fig. **30a**, **b**).

Fig. (30a). Canvas painting of Mahonia illustrated by Patanjali Research Foundation, Haridwar, India.

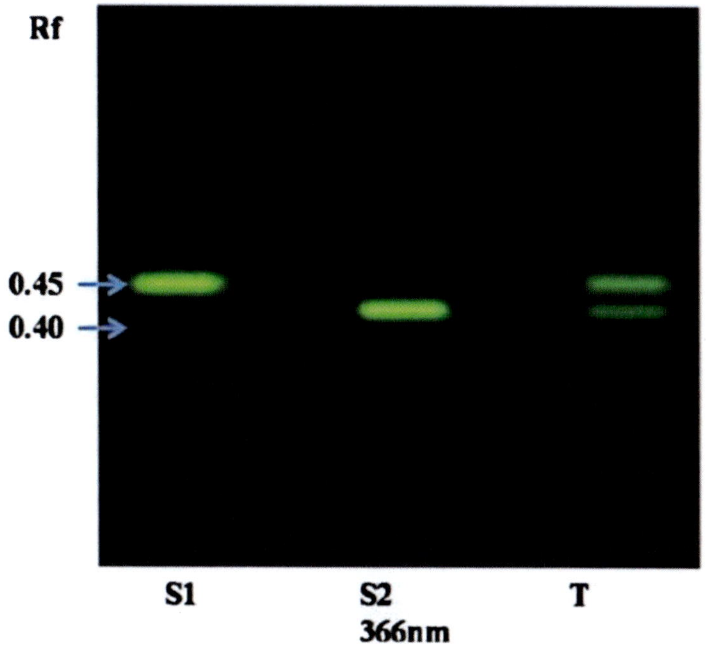

Fig. (30b). HPTLC fingerprint of Mahonia: S1: Berberine, S2: Palmatine, T1: Mahonia stem.

Boerhavia diffusa L. (Common Name: Punarnava)

Plant Part: Roots.

Solvent System: Toluene: ethyl acetate: formic acid: methanol (8: 8: 2: 2 v/v/v/v).

Adsorbent: TLC Silica gel $60F_{254}$ (Merck).

Standard Solution Preparation: 10 mg of boeravinone B was dissolved in 10 mL of methanol. 10μL of standard solution was applied on a TLC plate.

Sample Solution Preparation: 1000 mg of sample dissolved in 10 mL of methanol, sonicated and centrifuged. 6 μL of solution was applied on a TLC plate.

Medicinal Value: Used to cure disorders like intestinal colic, kidney diseases, skin problems and asthma (2).

Conclusion: HPTLC fingerprint confirms the presence of bio-active compound boeravinone B at 0.68 Rf in a sample similar to boeravinone B standard at 254 nm (Fig. **31a**, **b**).

Fig. (31a). Canvas painting of Purnarnava illustrated by Patanjali Research Foundation, Haridwar, India .

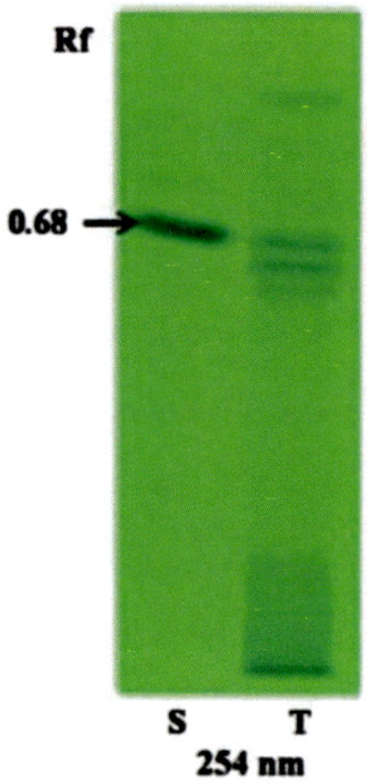

Fig. (31b). HPTLC fingerprint of Punarnava: S: Boeravinone B, T: Punarnava roots.

Brassica juncea (L.) Czern (Common Name: Red Mustard)

Plant Part: Seeds.

Solvent System: Toluene: ethyl acetate: formic acid (10: 10: 2 v/v/v).

Adsorbent: TLC Silica gel $60F_{254}$ (Merck).

Sample Solution Preparation: 500 mg sample dissolved in 10 mL of methanol, sonicated and centrifuged. 10 µL of solution was applied on a TLC plate.

Medicinal Value: It has fungicidal, antimicrobial and antioxidant activity (3).

Conclusion: Fingerprint of Red mustard shows a specific pattern of bands at 254 nm and 366 nm (Fig. **32a**, **b**).

Fig. (32a). Canvas painting of Red mustard illustrated by Patanjali Research Foundation, Haridwar, India.

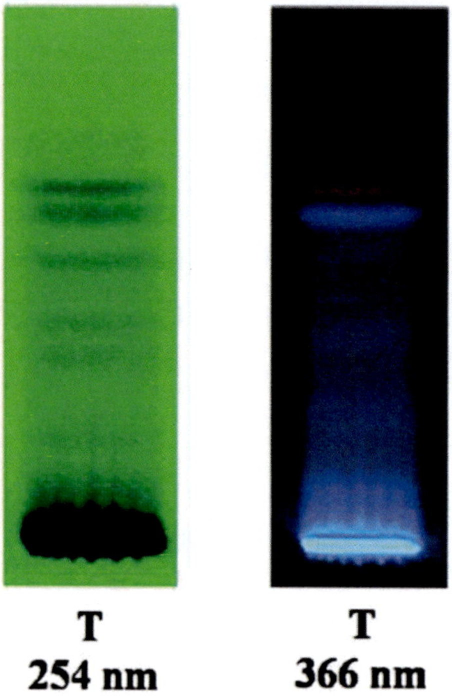

Fig. (32b). HPTLC fingerprint of Red mustard: T: Red mustard seeds.

Cannabis sativa L. (Common Name: Bhang)

Plant Part: Leaves.

Solvent System: Toluene: ethyl acetate: formic acid (6: 4: 0.3 v/v/v).

Adsorbent: TLC Silica gel $60F_{254}$ (Merck).

Sample Solution Preparation: 500 mg sample dissolved in 10 mL of ethanol, sonicated and centrifuged. 20 µL of solution was applied on a TLC plate.

Medicinal Value: Used as astringent, aphrodisiac, anodyne and sedative (3).

Conclusion: HPTLC fingerprint of Bhang displays a specific banding pattern of phytochemicals at 254 nm and red florescent bands at 366 nm (Fig. **33a**, **b**).

Fig. (33a). Canvas painting of Bhang illustrated by Patanjali Research Foundation, Haridwar, India.

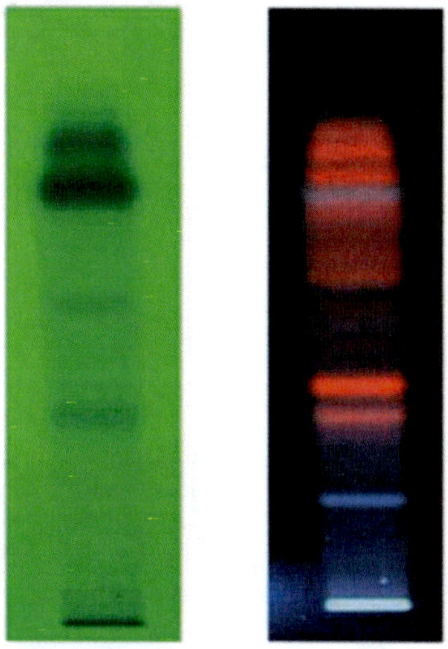

Fig. (33b). HPTLC fingerprint of Bhang: T: Bhang leaves.

Capsicum frutescens L. (Common Name: Chilli)

Plant Part: Fruit.

Solvent System: Toluene: ethyl acetate: formic acid (10: 10: 2 v/v/v).

Adsorbent: TLC Silica gel $60F_{254}$ (Merck).

Sample Solution Preparation: 500 mg sample dissolved in 10 mL of methanol, sonicated and centrifuged. 6 μL of solution was applied on a TLC plate.

Derivatizing Reagent: Anisaldehyde sulfuric acid.

Medicinal Value: It has antioxidant, antimicrobial and anti-inflammatory properties.

Conclusion: HPTLC fingerprint of Chilli shows a number of bands at 254 nm and under white light after derivatization (Fig. **34a**, **b**).

Fig. (34a). Canvas painting of Chilli illustrated by Patanjali Research Foundation, Haridwar, India.

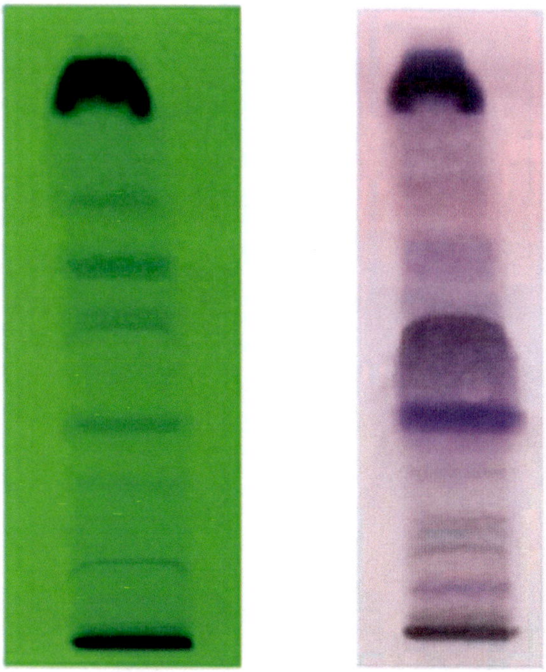

Fig. (34b). HPTLC fingerprint of Chilli: T: Chilli fruit.

Carica papaya L. (Common Name: Papaya)

Plant Part: Leaves.

Solvent System: Ethyl acetate: toluene: formic acid (5: 5: 0.5 v/v/v).

Adsorbent: TLC Silica gel $60F_{254}$ (Merck).

Sample Solution Preparation: 500 mg of sample dissolved in 10 mL methanol, sonicated and centrifuged. 8 μL of the solution was applied on a TLC plate.

Medicinal Value: Treat symptoms related to dengue fever. It has antibacterial, anti-inflammatory and anti-viral effects (2).

Conclusion: HPTLC fingerprint of Papaya shows specific banding pattern at 254 nm and 366 nm (Fig. **35a**, **b**).

Fig. (35a). Canvas painting of Papaya illustrated by Patanjali Research Foundation, Haridwar, India.

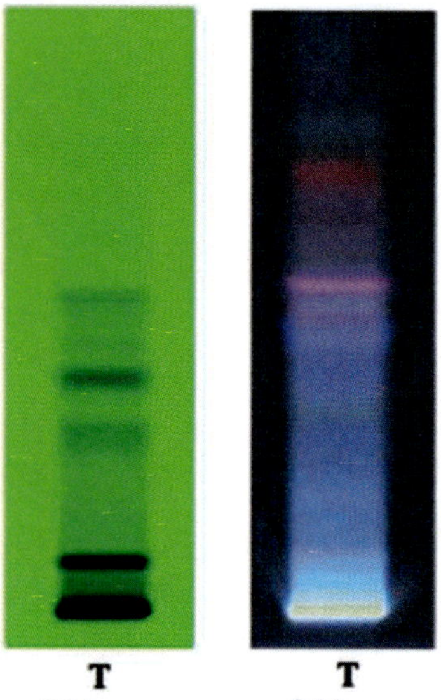

Fig. (35b). HPTLC fingerprint of Papaya: T: Papaya leaves.

Carissa carandas L. (Common Name: Karonda)

Plant Part: Fruit.

Solvent System: Ethyl acetate: toluene: formic acid (5: 5: 0.5 v/v/v).

Adsorbent: TLC Silica gel $60F_{254}$ (Merck).

Sample Solution Preparation: 500 mg of sample dissolved in 10 mL of methanol, sonicated and centrifuged. 2 µL of the solution was applied on a TLC plate.

Medicinal Value: Used to treat urinary tract infection, fever and blood disorder (1).

Conclusion: HPTLC fingerprint of Karonda shows a determined pattern of bands in both 254nm and 366 nm (Fig. **36a**, **b**).

Fig. (36a). Canvas painting of Karonda illustrated by Patanjali Research Foundation, Haridwar, India.

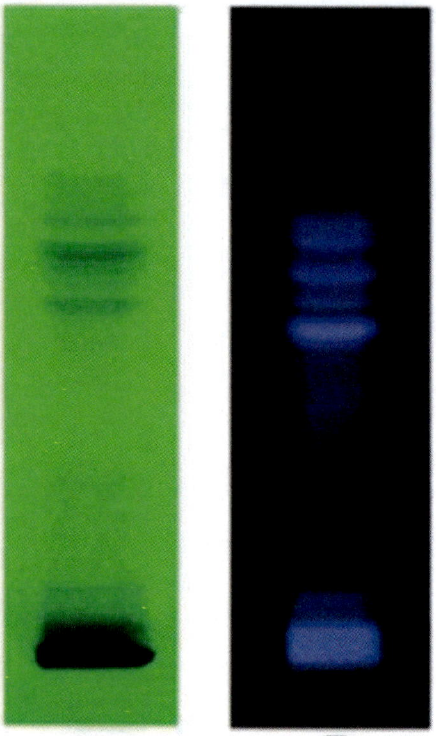

Fig. (36b). HPTLC fingerprint of Karonda: T: Karonda fruit.

Carum carvi L. (Common Name: Shah Jeera)

Plant Part: Seeds.

Solvent System: Ethyl acetate: acetic acid: water (10: 1: 2 v/v/v).

Adsorbent: TLC Silica gel $60F_{254}$ (Merck).

Sample Solution Preparation: 500 mg sample dissolved in 10 mL of methanol, sonicated and centrifuged. 10 µL of the solution was applied on a TLC plate.

Medicinal Value: Good for digestive health, respiratory system and skin diseases (2).

Conclusion: HPTLC fingerprint of Shah jeera shows a major fluorescent band at 366 nm and numbers of bands at 254 nm and 366 nm (Fig. **37a**, **b**).

Fig. (37a). Canvas painting of Shah Jeera illustrated by Patanjali Research Foundation, Haridwar, India.

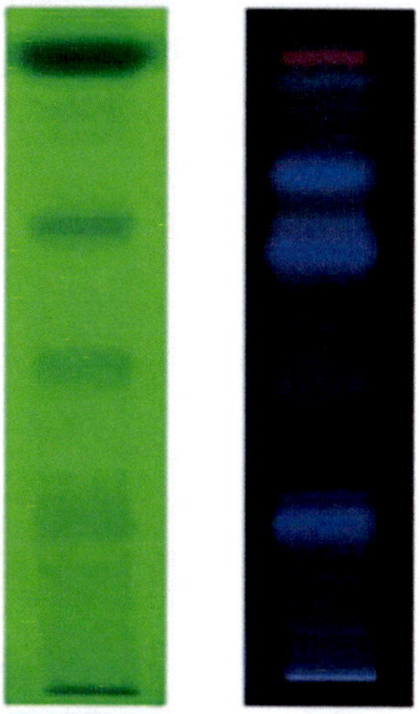

Fig. (37b). HPTLC fingerprint of Shah jeera: T: Shah jeera seeds.

Cassia fistula L. (Common Name: Amaltas)

Plant Part: Fruit.

Solvent System: Toluene: ethyl acetate (7: 3 v/v).

Adsorbent: TLC Silica gel $60F_{254}$ (Merck).

Sample Solution Preparation: 1000 mg of the sample dissolved in 10 mL of methanol, sonicated and centrifuged. 10 μL of the solution was applied on a TLC plate.

Derivatizing Reagent: Anisaldehyde sulfuric acid.

Medicinal Value: The plant has hepatoprotective, anti-inflammatory, antibacterial and wound healing properties.

Conclusion: HPTLC fingerprint of Amaltas shows major fluorescent bands at 366 nm, distributed at different Rf zones and purple colored bands in white light after derivatization (Fig. **38a**, **b**).

Fig. (38a). Canvas painting of Amaltas illustrated by Patanjali Research Foundation, Haridwar, India.

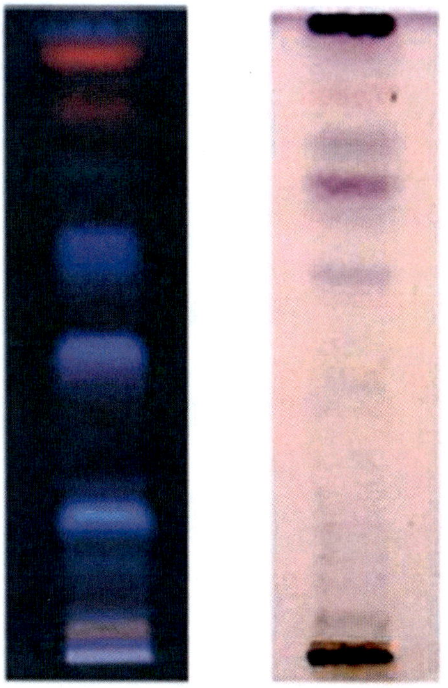

Fig. (38b). HPTLC fingerprint of Amaltas: T: Amaltas fruit.

Catharanthus roseus (L.) G.Don (Common Name: Sadabhar)

Plant Part: Leaves.

Solvent System: Toluene: ethyl acetate (7: 3 v/v).

Adsorbent: TLC Silica gel $60F_{254}$ (Merck).

Sample Solution Preparation: 1000 mg sample dissolved in 10 mL of methanol, sonicated and centrifuged. 10 µL of the solution was applied on a TLC plate.

Medicinal Value: The plant has hepatoprotective, anti-inflammatory, antibacterial and wound healing properties (3).

Conclusion: HPTLC fingerprint of Sadabhar shows major fluorescent bands at 366 nm, distributed at different Rf zones and purple colored bands in white light after derivatization (Fig. **39a**, **b**).

Fig. (39a). Canvas painting of Sadabhar illustrated by Patanjali Research Foundation, Haridwar, India.

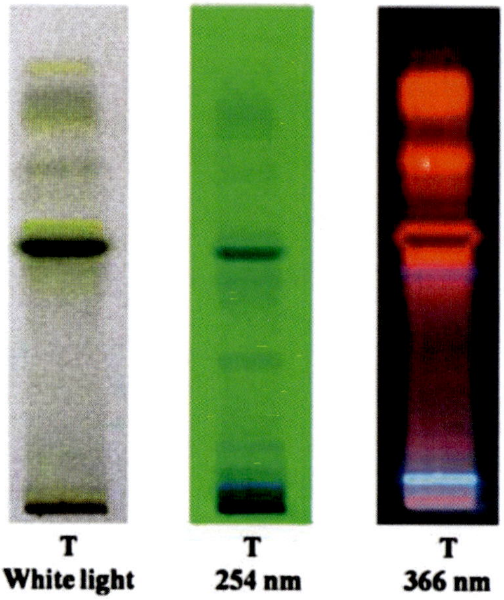

Fig. (39b). HPTLC fingerprinting of Sadabhar: T: Sadabhar leaves.

Cedrus deodara (Roxb. ex D.Don) G.Don (Common Name: Deodar)

Plant Part: Bark.

Solvent System: Ethyl acetate: toluene: formic acid (10: 10: 2 v/v/v).

Adsorbent: TLC Silica gel $60F_{254}$ (Merck).

Sample Solution Preparation: 1000 mg of sample dissolved in 10 mL of methanol, sonicated and centrifuged. 12 μL of the solution was applied on TLC plate.

Medicinal Value: Useful in curing piles, epilepsy, fever, stones in the kidney and bladder (2).

Conclusion: HPTLC fingerprint of Deodar shows a determined pattern of bands in both 254nm and 366 nm (Fig. **40a, b**).

Fig. (40a). Canvas painting of Deordar illustrated by Patanjali Research Foundation, Haridwar, India.

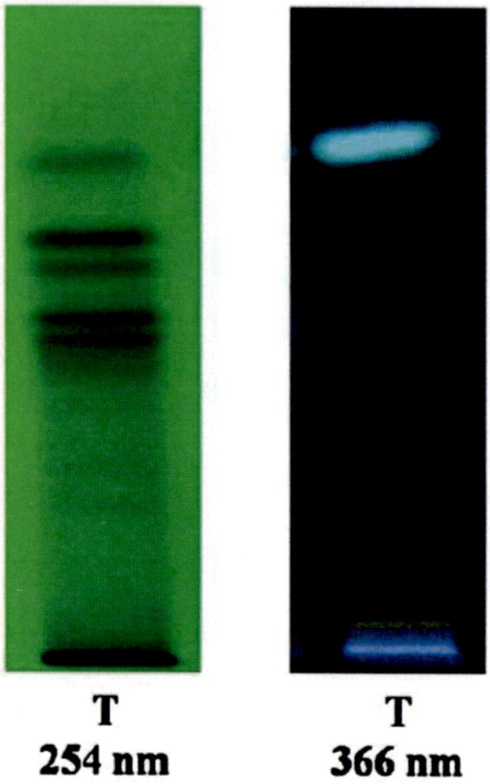

Fig. (40b). HPTLC fingerprinting of Deodar: T: Deodar Bark.

Celastrus paniculatus Willd (Common Name: Malkangani)

Plant Part: Seeds.

Solvent System: Chloroform: methanol (9.5: 0.5 v/v).

Adsorbent: TLC Silica gel $60F_{254}$ (Merck).

Sample Solution Preparation: 500 mg of sample dissolved in 10 mL of methanol, sonicated and centrifuged. 6 µL of the solution was applied on a TLC plate.

Medicinal Value: Used to improve memory power. It has anti-inflammatory, gastro protective, and wound healing properties (3).

Conclusion: HPTLC fingerprint of Malkangani shows some light bands at 254 nm and strong fluorescent bands under 366 nm (Fig. **41a**, **b**).

Fig. (41a). Canvas painting of Malkangani illustrated by Patanjali Research Foundation, Haridwar, India.

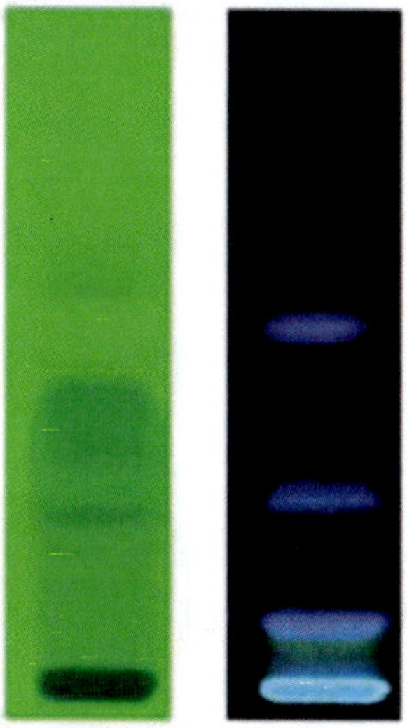

Fig. (41b). HPTLC fingerprinting of Malkangani: T: Malkangani seeds.

Chlorophytum borivilianum **Santapau & R.R.Fern (Common Name: Safed musli)**

Plant Part: Root.

Solvent System: Toluene: ethyl acetate: formic acid: methanol (8: 8: 2: 2 v/v/v/v).

Adsorbent: TLC Silica gel $60F_{254}$ (Merck)

Sample Solution Preparation: 1000 mg sample dissolved in 10 mL of methanol, sonicated and centrifuged. 10 μL of the solution was applied on a TLC plate.

Medicinal Value: Relieves excessive thirst, rejuvenates the whole body, treats urinary tract disorders, skin disorders and improves muscle strength.

Conclusion: Fingerprint of Safed musli shows a specific pattern of bands at 254 nm and a major fluorescent band at 366 nm (Fig. **42a**, **b**).

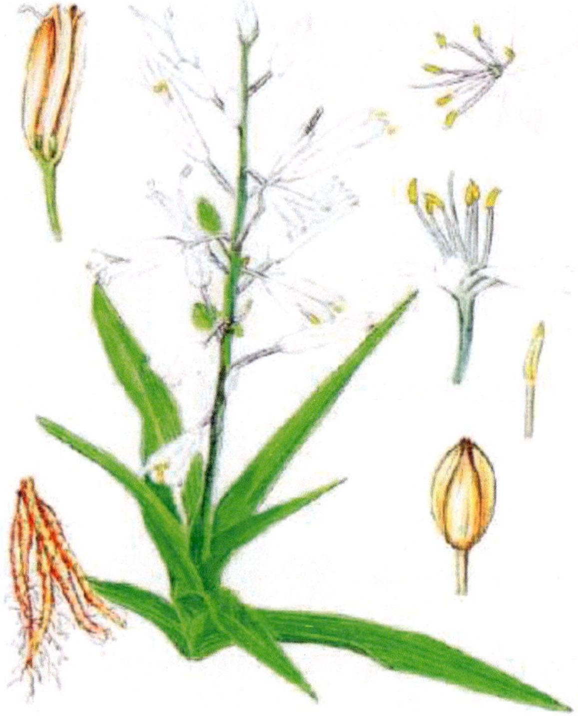

Fig. (42a). Canvas painting of Safed musli illustrated by Patanjali Research Foundation, Haridwar, India.

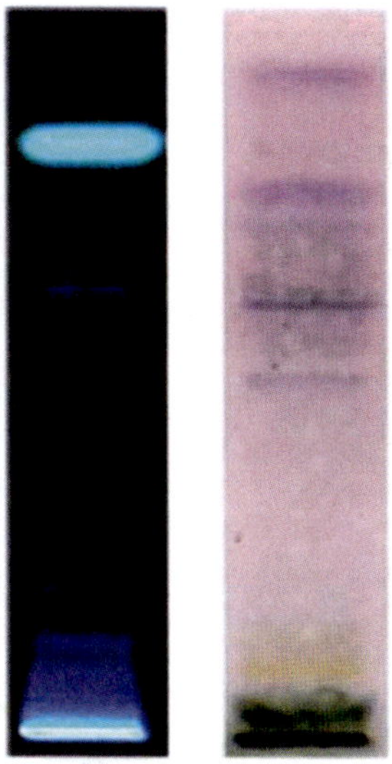

Fig. (42b). HPTLC fingerprinting of Safed musli: T: Safed musli root.

Cinnamomum camphora (L.) J. Presl (Common Name: Camphor)

Plant Part: Leaves.

Solvent System: Hexane: ethyl acetate (8: 2 v/v).

Adsorbent: TLC Silica gel $60F_{254}$ (Merck).

Sample Solution Preparation: 500 mg of the sample dissolved in 10 mL of hexane, sonicated and centrifuged. 6 μL of the solution was applied on a TLC plate.

Derivatization: Anisaldehyde sulfuric acid.

Medicinal Value: Used for the treatment of cardiovascular disorders, cold and fevers, respiratory diseases (1).

Conclusion: HPTLC fingerprint of Camphor numbers of bands at 254 nm, 366nm and in white light after derivatization. Red colored fluorescent bands and purple colored bands appear at 366 nm and in white light, respectively (Fig. **43a**, **b**).

Fig. (43a). Canvas painting of Camphor illustrated by Patanjali Research Foundation, Haridwar, India.

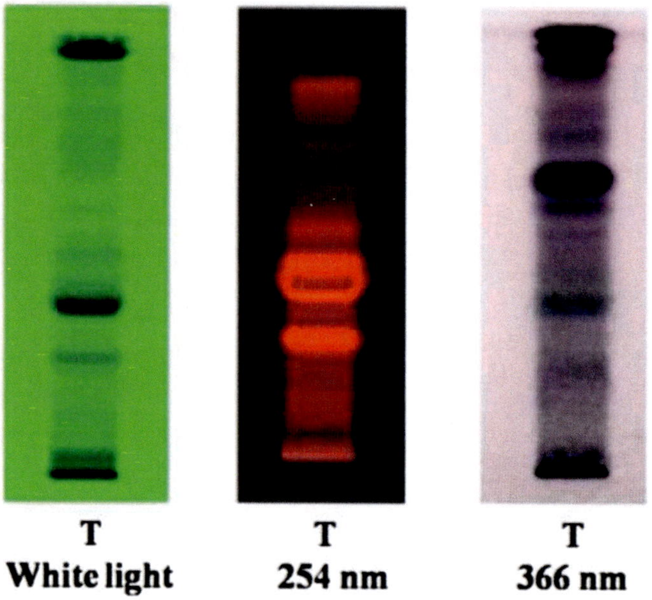

Fig. (43b). HPTLC fingerprinting of Camphor: T: Camphor leaves.

Cinnamomum tamala (Buch.-Ham.) T.Nees & C.H.Eberm (Common Name: Bay Leaves)

Plant Part: Leaves.

Solvent System: Toluene: ethyl acetate: formic acid (10: 10: 2 v/v/v).

Adsorbent: The TLC silica gel $60F_{254}$ (Merck).

Sample Solution Preparation: 1000 mg sample dissolved in 10 mL of methanol, sonicated and centrifuged. 6 µL of the solution was applied on a TLC plate.

Derivatization: Anisaldehyde sulfuric acid.

Medicinal Value: Used in cold, cough, asthma and abdominal pain (2).

Conclusion: Specific patterns of bands observed at 254 nm, 366 nm and under white light after derivatization of Bay leaves fingerprint (Fig. **44a**, **b**).

Fig. (44a). Canvas painting of Bay leaves illustrated by Patanjali Research Foundation, Haridwar, India.

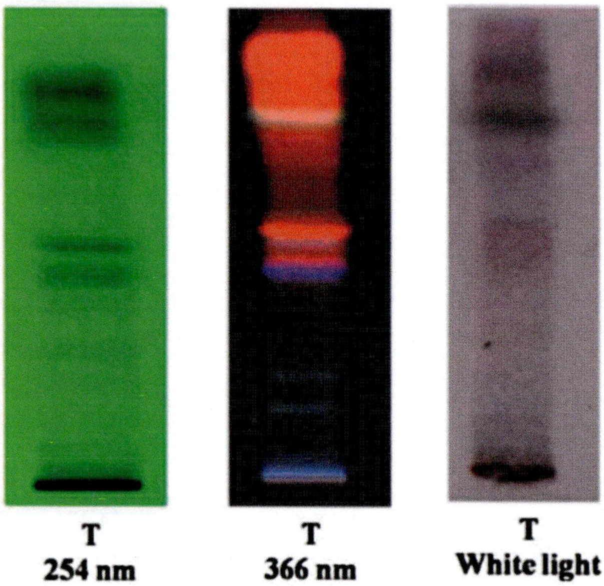

Fig. (44b). HPTLC fingerprinting of Bay leaves: T: Bay leaves.

Cinnamomum verum J. Presl (Common Name: Cinnamon)

T1: Cinnamon bark alcoholic extract, T2: Cinnamon bark water extract.

Plant Part: Bark.

Solvent System: Chloroform: ethyl acetate: acetic acid (7: 3: 1 v/v/v).

Adsorbent: TLC Silica gel $60F_{254}$ (Merck).

Standard Solution Preparation: 10 mg of cinnamic acid was dissolved in 10 mL of methanol. 10 µL of standard solution was applied on a TLC plate.

Sample Solution Preparation: 100 mg of each sample dissolved in 10 mL of their respective solvents, sonicated and centrifuged. 20 µL of the solution was applied on a TLC plate.

Medicinal Value: Useful in asthma, cough, cold, sore throat and indigestion (2).

Conclusion: Photograph of TLC attesting the presence of marker cinnamic acid in Cinnamon. Rf value 0.53 confirms cinnamic acid in both the samples (Fig. **45a**, **b**).

Fig. (45a). Canvas painting of Cinnamon illustrated by Patanjali Research Foundation, Haridwar, India.

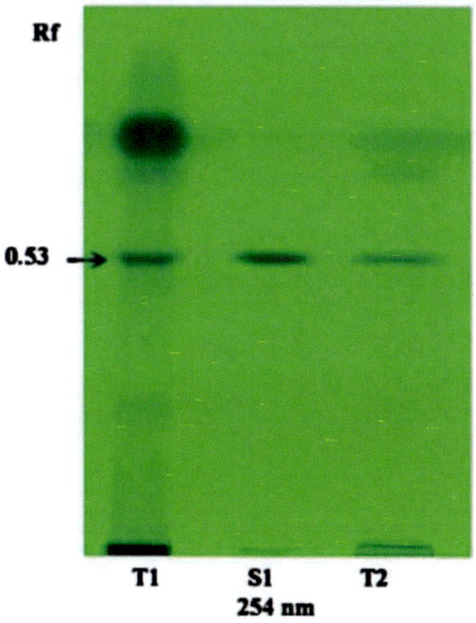

Fig. (45b). HPTLC fingerprinting of Cinnamon: S1: Cinnamic acid,.

Cissus quadrangularis L. (Common Name: Hadjod)

Plant Part: Stem.

Solvent System: Toluene: ethyl acetate: formic acid (5: 5: 1 v/v/v).

Adsorbent: TLC Silica gel $60F_{254}$ (Merck).

Sample Solution Preparation: 1000 mg sample dissolved in 10 mL of their respective solvents, sonicated and centrifuged. 20 μL of the solution was applied on a TLC plate.

Medicinal Value: Useful in strengthening the bones, and healing fractures and has anti-inflammatory properties (3).

Conclusion: HPTLC fingerprint of Hadjod shows specific bands at 254 nm, 366 nm and in white light after derivatization (Fig. **46a**, **b**).

Fig. (46a). Canvas painting of Hadjod illustrated by Patanjali Research Foundation, Haridwar, India.

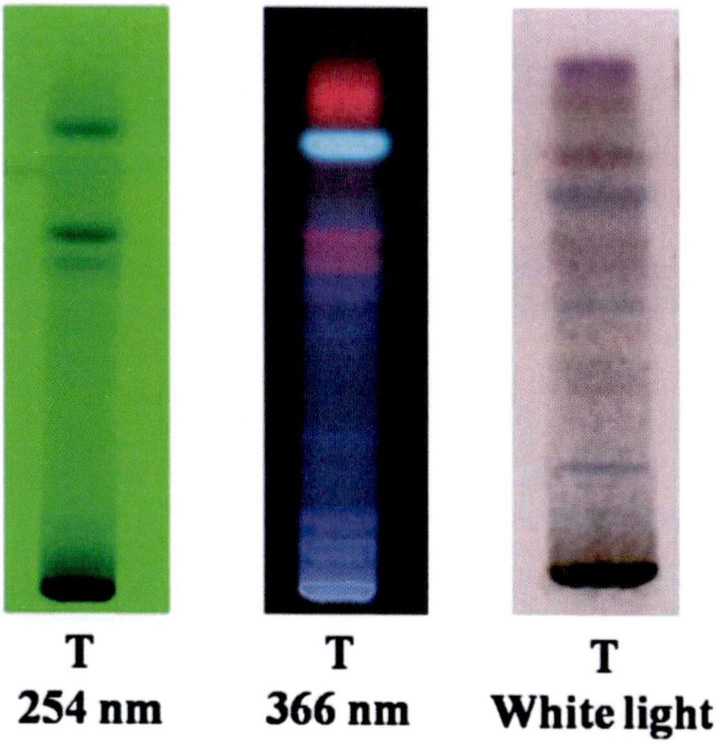

Fig. (46b). HPTLC fingerprinting of Hadjod: T: Hadjod stem.

Cocculus hirsutus **(L.) W.Theob (Common Name: Jalamini)**

Plant Part: Leaves.

Solvent System: Toluene: ethyl acetate: formic acid (10: 10: 2 v/v/v).

Adsorbent: TLC Silica gel $60F_{254}$ (Merck).

Sample Solution Preparation: 500 mg sample dissolved in 10 mL of methanol, sonicated and centrifuged. 20 μL of the solution was applied on a TLC plate.

Medicinal Value: Used for the treatment of fever, skin diseases, stomach and urinary disorders.

Conclusion: HPTLC fingerprint of Jaljamini displays a specific banding pattern of phytochemicals in white light before derivatization, at 254 nm and 366 nm (Fig. **47a**, **b**).

Fig. (47a). Canvas painting of Jaljamini illustrated by Patanjali Research Foundation, Haridwar, India.

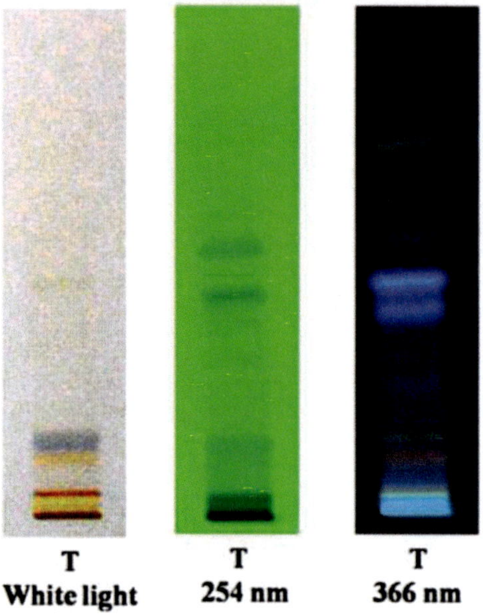

Fig. (47b). HPTLC fingerprinting of Jaljamini: T: Jaljamini leaves.

Commiphora mukul **(Hook. ex Stocks) Engl (Common Name: Guggul)**

Plant Part: Resin.

Solvent System: Ethyl acetate: toluene: formic acid (10: 9: 1 v/v/v).

Adsorbent: TLC Silica gel $60F_{254}$ (Merck).

Standard Solution Preparation: 10 mg of Guggulsterone Z was dissolved in 10 mL of methanol. 2 µL of the standard solution was applied on a TLC plate.

Sample Solution Preparation: 500 mg of sample dissolved in 10 mL methanol, sonicated and centrifuged. 6 µL of the solution was applied on a TLC plate.

Medicinal Value: Useful in cough, sore throat, asthma and bronchitis (1).

Conclusion: Rf 0.76 of the sample confirmed the presence of guggulsterone Z in Guggul by HPTLC fingerprint (Fig. **48a**, **b**).

Fig. (48a). Canvas painting of Guggul illustrated by Patanjali Research Foundation, Haridwar, India.

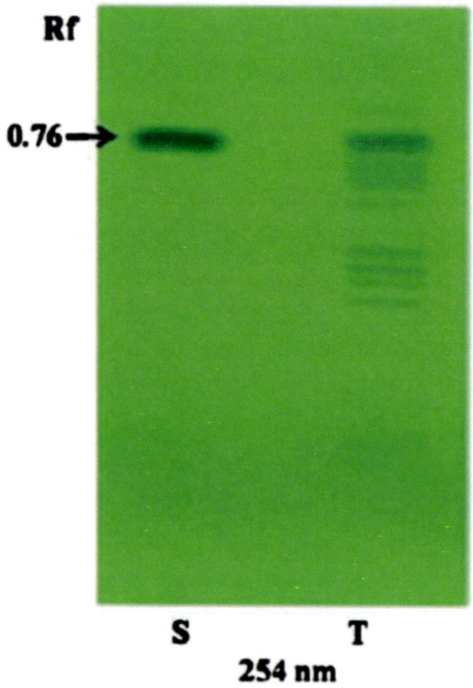

Fig. (48b). HPTLC fingerprinting of Guggul: S: Guggulsterone Z, T: Guggul resin.

Convolvulus prostratus Forssk (Common Name: Shankhpushpi)

Plant Part: Whole plant.

Solvent System: Toluene: Ethyl acetate: Formic acid (6: 4: 0.3 v/v/v).

Adsorbent: TLC Silica gel $60F_{254}$ (Merck).

Sample Solution Preparation: 500 mg sample dissolved in 10 mL alcohol, sonicated and centrifuged. 20 µL of the solution was applied on a TLC plate.

Medicinal Value: Useful in skin diseases, boosts memory and reduces inflammation (3).

Conclusion: HPTLC image of Shankhpushpi showing a major band at 254 nm and numbers of band at 366 nm (Fig. **49a**, **b**).

Fig. (49a). Canvas painting of Shankhpushpi illustrated by Patanjali Research Foundation, Haridwar, India.

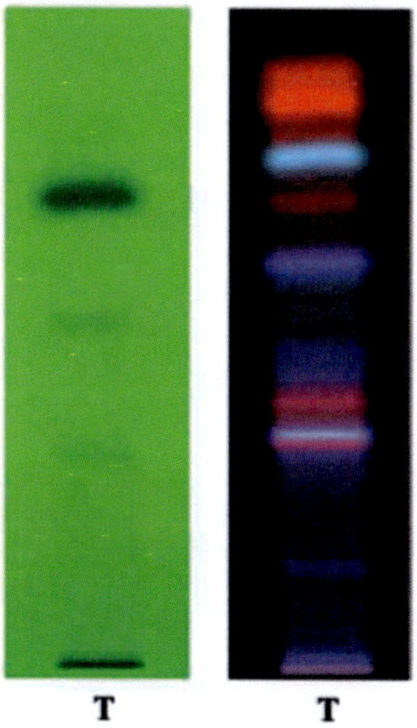

T
254 nm

T
366 nm

Fig. (49b). HPTLC fingerprinting of Shankhpushpi: T: Shankhpushpi whole plant.

Cordia dichotoma G.Forst (Common Name: Lisoda)

Plant Part: Fruit.

Solvent System: Hexane: ethyl acetate (8: 2 v/v).

Adsorbent: TLC Silica gel $60F_{254}$ (Merck).

Standard Solution Preparation: 10 mg of Beta -sitosterol was dissolved in 10 mL of methanol. 10 µL of the standard solution was applied on TLC plate.

Sample Solution Preparation: 500 mg of the sample dissolved in 10 mL hexane, sonicated and centrifuged. 20 µL of the solution was applied on a TLC plate.

Derivatization: Anisaldehyde sulfuric acid.

Medicinal Value: Useful to treat diarrhoea, irritable bowel syndrome and intestinal worms (3).

Conclusion: HPTLC fingerprint confirms the presence of beta-sitosterol in sample at Rf 0.58 with respect to standard. Purple colored band of beta-sitosterol, observed after derivatization under white light (Fig. **50a**, **b**).

Fig. (50a). Canvas painting of Lisoda illustrated by Patanjali Research Foundation, Haridwar, India.

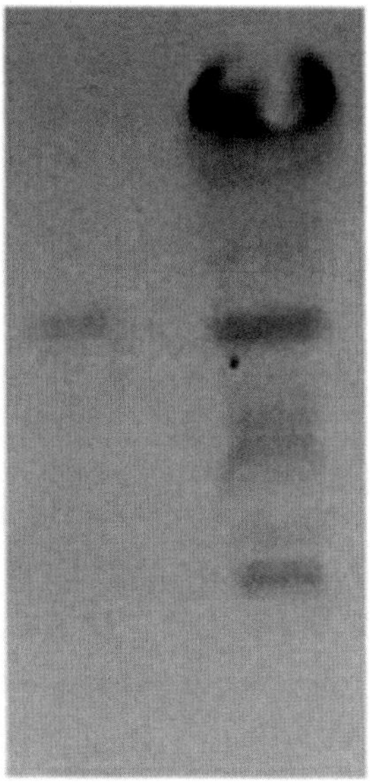

Fig. (50b). HPTLC fingerprinting of Lisoda: S: Beta-sitosterol, T: Lisoda fruit.

Coriandrum sativum L. (Common Name: Coriander)

Plant Part: Seeds.

Solvent System: Toluene: ethyl acetate: formic acid (10: 10: 1 v/v/v).

Adsorbent: TLC Silica gel $60F_{254}$ (Merck).

Sample Solution Preparation: 1000 mg sample dissolved in 10 mL of methanol, sonicated and centrifuged. 10 µL of the solution was applied on a TLC plate.

Medicinal Value: Good for digestive system, reduces inflammation, treats wound and mouth ulcers (3).

Conclusion: HPTLC fingerprint of Coriander shows a major fluorescent band at 366 nm and numbers of bands at 254 nm and under white light after derivatization (Fig. **51a**, **b**).

Fig. (51a). Canvas painting of Coriander illustrated by Patanjali Research Foundation, Haridwar, India.

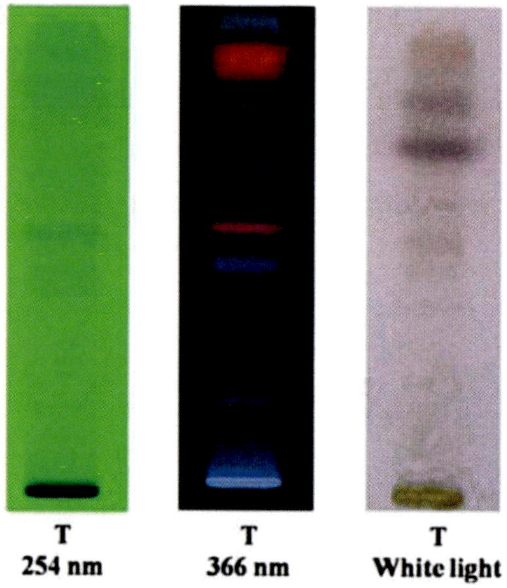

Fig. (51b). HPTLC fingerprinting of Coriander: T: Coriander seeds.

Cornus capitata Wall (Common Name: Thanboi)

Plant Part: Leaves.

Solvent System: Ethyl acetate: toluene: formic acid (24: 16: 2 v/v/v).

Adsorbent: TLC Silica gel $60F_{254}$ (Merck).

Standard Solution Preparation: 10 mg of gallic acid was dissolved in 10 mL of methanol. 10 μL of standard solution was applied on a TLC plate.

Sample Solution Preparation: 500 mg sample dissolved in 10 mL methanol sonicated and centrifuged. 4 μL of the solution was applied on a TLC plate.

Medicinal Value: It has antimicrobial, antioxidant and anti-inflammatory properties.

Conclusion: HPTLC image confirms the identification of bio-active marker gallic acid at Rf 0.41, with respect to gallic acid standard (Fig. **52a**, **b**).

Fig. (52a). Canvas painting of Thanboi illustrated by Patanjali Research Foundation, Haridwar, India.

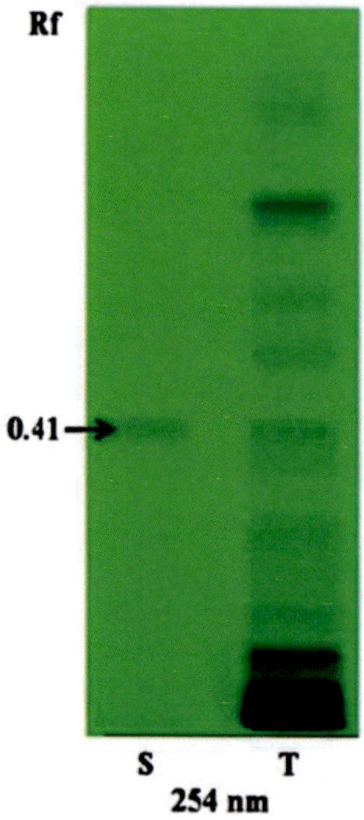

Fig. (52b). HPTLC fingerprinting of Thanboi: S: Gallic acid, T: Thanboi leaves.

Cressa cretica L. (Common Name: Rudanti)

Plant Part: Fruit.

Solvent System: Toluene: ethyl acetate (7: 3 v/v).

Adsorbent: TLC Silica gel $60F_{254}$ (Merck).

Sample Solution Preparation: 500 mg of sample dissolved in 10 mL water: methanol (70: 30), sonicated and centrifuged. 8µL of the solution was applied on a TLC plate.

Medicinal Value: Used to treat respiratory diseases like cough, bronchial asthma and also useful in tuberculosis (3).

Conclusion: HPTLC fingerprint of Rudanti shows fluorescent bands at 366 nm and blue colored bands under white light after derivatization (Fig. **53a**, **b**).

Fig. (53a). Canvas painting of Rudanti illustrated by Patanjali Research Foundation, Haridwar, India.

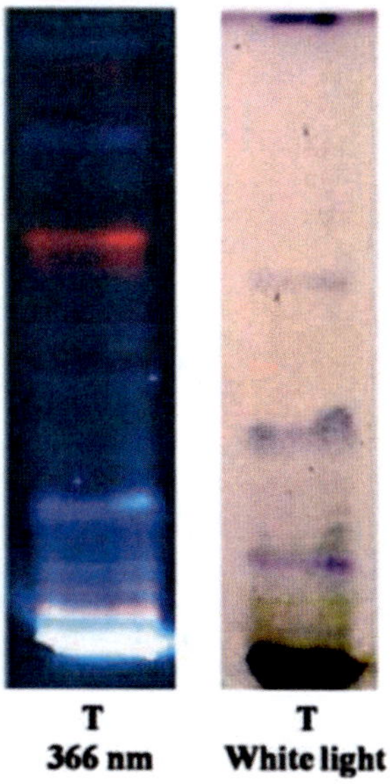

Fig. (53b). HPTLC fingerprinting of Rudanti: T: Rudanti fruit.

Crinum latifolium L. (Common Name: Sudarshan)

Plant Part: Leaves.

Solvent System: Chloroform: methanol (8: 2 v/v):

Adsorbent: TLC Silica gel $60F_{254}$ (Merck)

Sample Solution Preparation: 500 mg of sample dissolved in 10 mL water: methanol (70: 30 v/v) sonicated and centrifuged. 10 μL of the solution was applied on TLC plate.

Medicinal Value: Useful in ear infection, skin problems, arthritis; it is anthelminthic (3).

Conclusion: Specific patterns of phytochemicals were observed in 254 nm and blue fluorescent bands were observed under 366 nm in HPTLC fingerprint of Sudarshan (Fig. **54a**, **b**).

Fig. (54a). Canvas painting of Sudarshan illustrated by Patanjali Research Foundation, Haridwar, India.

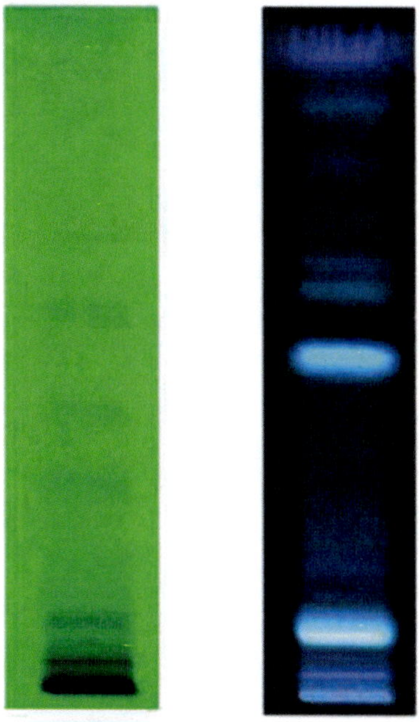

Fig. (54b). HPTLC fingerprinting of Sudarshan, T: Sudarshan leaves.

Cullen corylifolium (L.) Medik (Common Name: Bakuchi)

Plant Part: Seeds.

Solvent System: Ethyl acetate: toluene: formic acid (5: 2: 1.5 v/v/v).

Adsorbent: TLC Silica gel $60F_{254}$ (Merck).

Standard Solution Preparation: 10 mg of psoralen was dissolved in 10 mL of methanol. 1 µL of standard solution was applied on a TLC plate.

Sample Solution Preparation: 500 mg of the sample dissolved in 10 mL methanol sonicated and centrifuged. 8 µL of the solution was applied on a TLC plate.

Medicinal Value: Used to treat various skin problems, and promote hair growth and digestion (2).

Conclusion: HPTLC fingerprint of bakuchi matches with reference standard, showing specific band of bio-active marker psoralen at 0.61 Rf (Fig. **55a**, **b**).

Fig. (55a). Canvas painting of Bakuchi illustrated by Patanjali Research Foundation, Haridwar, India.

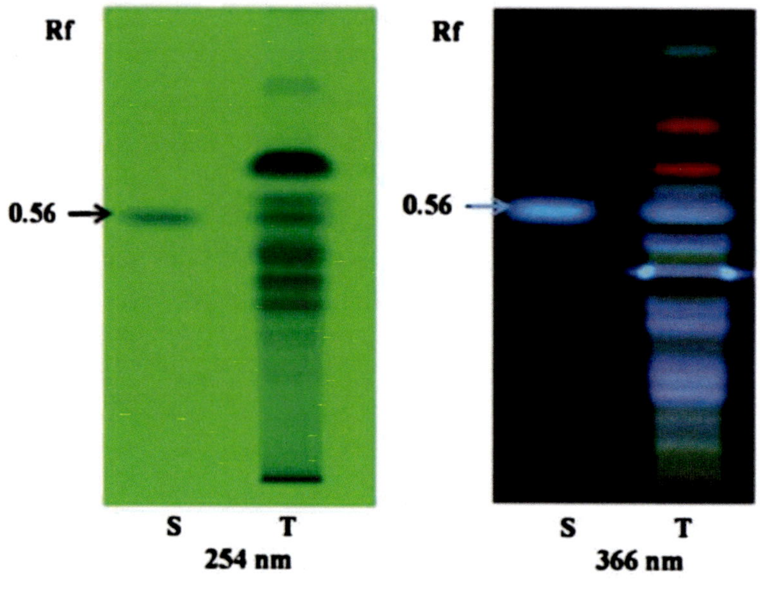

Fig. (55b). HPTLC fingerprint of Bakuchi: S: Psoralen, T: Bakuchi seeds.

Cuminum cyminum L. (Common Name: Cumin)

Plant Part: Seeds.

Solvent System: Ethyl acetate: toluene: formic acid (10: 10: 1 v/v/v).

Adsorbent: TLC Silica gel $60F_{254}$ (Merck).

Sample Solution Preparation: 500 mg sample dissolved in 10 mL of methanol, sonicated and centrifuged. 8 μL of the solution was applied on a TLC plate.

Medicinal Value: Useful in skin and acne problem, improves digestion, improves cardiac health and boost immunity (2).

Conclusion: HPTLC fingerprint of Cumin captured at 254 nm, 366 nm showing bands of phytochemicals at 254 nm and fluorescent bands at 366 nm (Fig. **56a**, **b**).

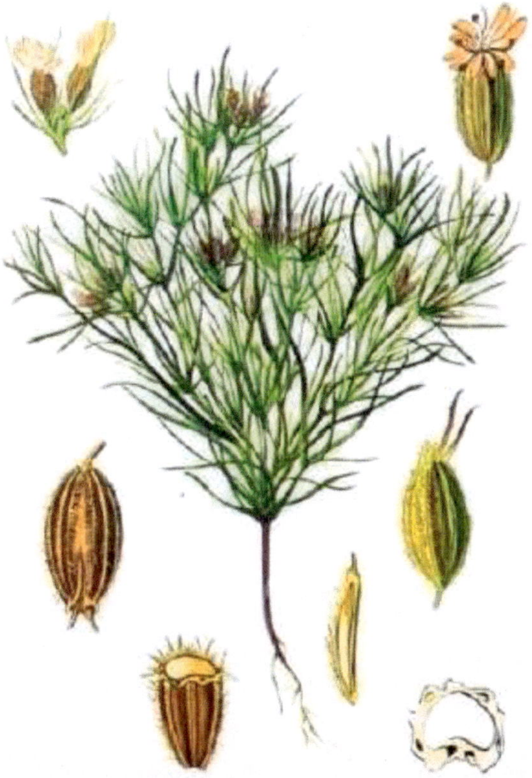

Fig. (56a). Canvas painting of Cumin illustrated by Patanjali Research Foundation, Haridwar, India.

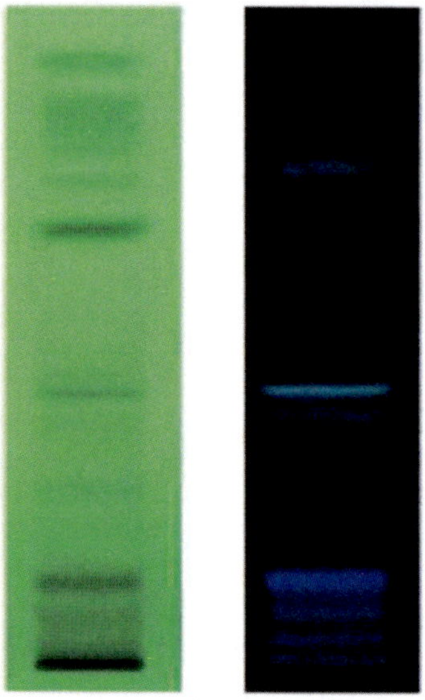

Fig. (56b). HPTLC fingerprint of Cumin: T: Cumin seeds.

Curcuma longa L. (Common Name: Turmeric)

Plant Part: Rhizome.

Solvent System: Toluene: acetic acid (4: 1 v/v).

Adsorbent: TLC Silica gel $60F_{254}$ (Merck).

Standard Solution Preparation: 10 mg of each curcumenoids was dissolved in 10 mL of methanol. 2 µL of the standard solution was applied on a TLC plate.

Sample Solution Preparation: 500 mg of the sample dissolved in 10 mL of methanol, sonicated and centrifuged. 6 µL of the solution was applied on a TLC plate.

Medicinal Value: Useful in cough, and skin diseases and improves liver function. It has antioxidant and anti-inflammatory properties (3).

Conclusion: HPTLC fingerprint validates the presence of signature marker components in Turmeric. Samples showing the bands of marker compounds at 0.27 (Bisdemethoxycurcumin), 0.38 (Demethoxycurcumin) and 0.53 (Curcumin) Rf with respect to the standard compounds (Fig. **57a**, **b**).

Fig. (57a). Canvas painting of Turmeric illustrated by Patanjali Research Foundation, Haridwar, India.

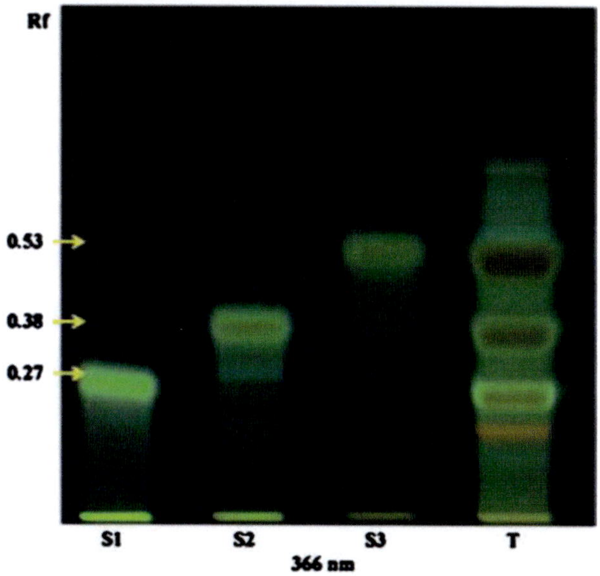

Fig. (57b). HPTLC fingerprint of Turmeric: S1: Bisdemethoxycurcumin, S2: Demethoxycurcumin, S3: Curcumin, T: Turmeric rhizome.

Cyperus scariosus R. Br (Common Name: Nagarmotha)

Plant Part: Rhizome.

Solvent System: Ethyl acetate: toluene: formic acid (5: 5: 0.5 v/v/v).

Adsorbent: TLC Silica gel $60F_{254}$ (Merck).

Sample Solution Preparation: 500 mg of the sample dissolved in 10 mL methanol, sonicated and centrifuged. 12 μL of the solution was applied on TLC plate.

Medicinal Value: Useful in indigestion, abdominal pain and diarrhoea.

Conclusion: HPTLC fingerprint of Nagarmotha shows specific binding pattern at 254 nm and 366 nm (Fig. **58a, b**).

Fig. (58a). Canvas painting of Nagarmotha illustrated by Patanjali Research Foundation, Haridwar, India.

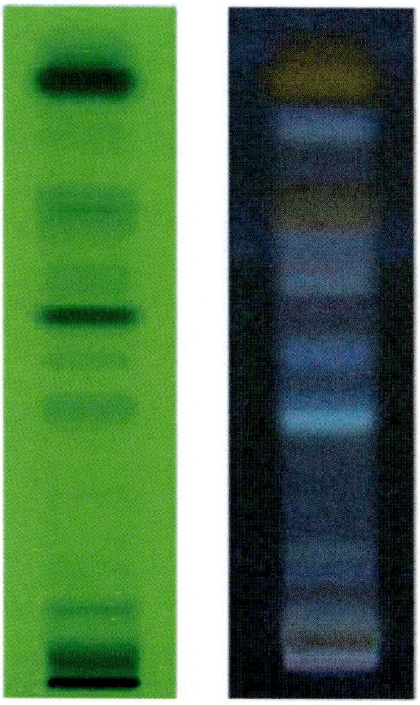

Fig. (58b). HPTLC fingerprint of Nagarmotha: T: Nagarmotha rhizome.

Datura stramonium L. (Common Name: Datura)

Plant Part: Stem.

Solvent System: Toluene: ethyl acetate: formic acid (10: 10: 1 v/v/v).

Adsorbent: TLC Silica gel $60F_{254}$ (Merck).

Sample Solution Preparation: 1000 mg of the sample dissolved in 10 mL hexane, sonicated and centrifuged. 20 µL of the solution was applied on a TLC plate.

Derivatization: Anisaldehyde sulfuric acid.

Medicinal Value: Helps in digestion, treats skin disorders and stimulates appetite (2).

Conclusion: HPTLC fingerprint represents bands of phytochemicals at 254 nm, fluorescent bands at 366 nm and blue, bluish violet bands under white light after derivatization in Datura (Fig. **59a**, **b**).

Fig. (59a). Canvas painting of Datura illustrated by Patanjali Research Foundation, Haridwar, India.

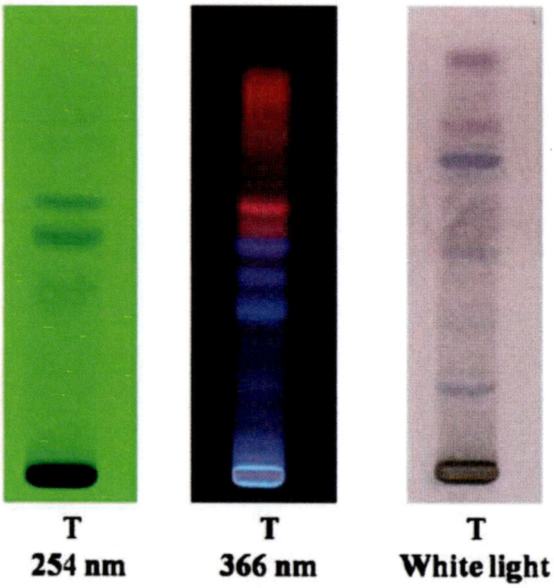

Fig. (59b). HTPLC fingerprint of Datura: T: Datura stem.

Daucus carota L. (Common Name: Carrot)

Plant Part: Root.

Solvent System: Toluene: ethyl acetate: formic acid (10: 10: 1 v/v/v).

Adsorbent: TLC Silica gel $60F_{254}$ (Merck).

Sample Solution Preparation: 500 mg of the sample dissolved in 10 mL water: methanol (70: 30 v/v) sonicated and centrifuged. 10 µL of the solution was applied on TLC plate.

Medicinal Value: Used as diuretic; improves vision, digestion and immunity (1).

Conclusion: Specific pattern of phytochemicals observed at 254 nm; blue and red fluorescent bands were observed under 366 nm of Carrot fingerprint (Fig. **60a**, **b**).

Fig. (60a). Canvas painting of Carrot illustrated by Patanjali Research Foundation, Haridwar, India.

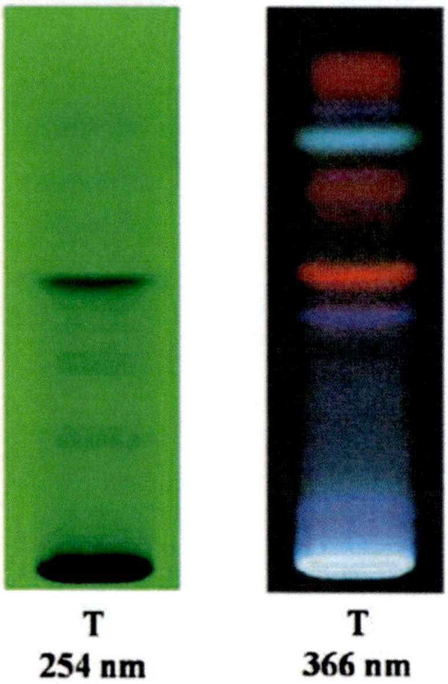

Fig. (60b). HPTLC fingerprint of Carrot: T: Carrot root.

Diplocyclos palmatus (L.) C.Jeffrey (Common Name: Shivlingi)

Plant Part: Seeds.

Solvent System: Chloroform: methanol (9.5: 0.5 v/v).

Adsorbent: TLC Silica gel $60F_{254}$ (Merck).

Standard Solution Preparation: 10 mg of Beta sitosterol was dissolved in 10 mL of methanol. 12 µL of the standard solution was applied on a TLC plate.

Sample Solution Preparation: 500 mg of the sample dissolved in 10 mL methanol, sonicated and centrifuged. 8 µL of the solution was applied on a TLC plate.

Derivatization: Anisaldehyde sulfuric acid.

Medicinal Value: It is fertility booster, spermatogenic, antipyretic, and antimicrobial (3).

Conclusion: HPTLC fingerprint of sample matches with reference standard, showing a specific band of bio-active marker beta-sitosterol at 0.58 Rf (Fig. **61a**, **b**).

Fig. (61a). Canvas painting of Shivlingi illustrated by Patanjali Research Foundation, Haridwar, India.

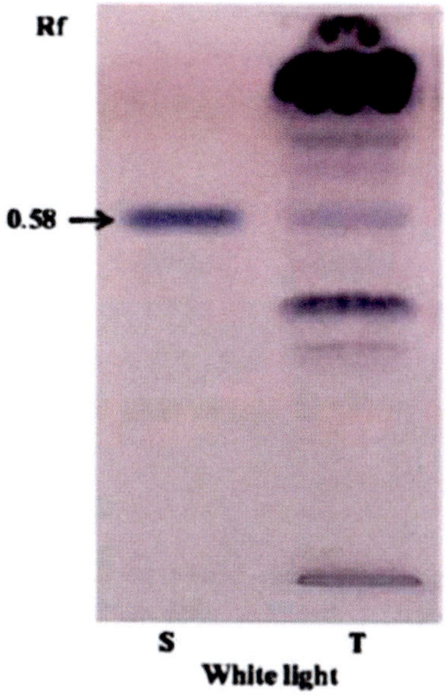

Fig. (61b). HPTLC fingerprint of Shivlingi: S: Beta-sitosterol, T: Shivling seeds.

Eclipta prostrata (L.) L (Common Name: Bhringraj)

Plant Part: Whole plant.

Solvent System: Ethyl acetate: toluene: formic acid (10: 10: 2 v/v/v).

Adsorbent: TLC Silica gel $60F_{254}$ (Merck).

Sample Solution Preparation: 1000 mg sample dissolved in 10 mL methanol sonicated and centrifuged. 15 μL of the solution was applied on a TLC plate.

Medicinal Value: Promotes hair growth, prevents hair fall, treats dandruff, dry scalp, and scalp infection (3).

Conclusion: HPTLC fingerprint of Bhringraj shows blue and green colored fluorescent bands at 366 nm and blue and purple bands under white light after derivatization (Fig. **62a**, **b**).

Fig. (62a). Canvas painting of Bhringraj illustrated by Patanjali Research Foundation, Haridwar, India.

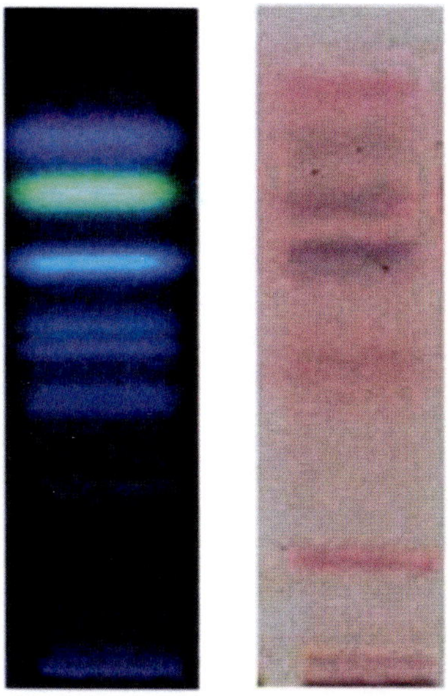

Fig. (62b). HPTLC fingerprint of Bhringraj: T: Bhringraj whole plant.

Elaeocarpus serratus L. (Common Name: Rudraksh)

Plant Part: Seeds.

Solvent System: Dichloromethane: methanol: acetic acid (10: 0.2: 0.05 v/v/v).

Adsorbent: TLC Silica gel $60F_{254}$ (Merck).

Standard Solution Preparation: 10 mg of Vanillin was dissolved in 10 mL of methanol. 20 μL of standard was applied on TLC plate.

Sample Solution Preparation: 1000 mg SCFE oil was dissolved in 10 mL petroleum ether and transferred to a separating funnel. For phase separation, 10 mL water was added and shaken well. Water layer was collected and dried.It was then reconstituted in 2 mL of methanol and 10 μL was applied on a TLC plate.

Medicinal Value: Used to cure headache, fever, and skin disease, and improve memory. (3).

Conclusion: HPTLC fingerprint of Rudraksh confirms the presence of vanillin at 0.54 Rf with respect to vanillin standard (Fig. **63a**, **b**).

Fig. (63a). Canvas painting of Rudraksh illustrated by Patanjali Research Foundation, Haridwar, India.

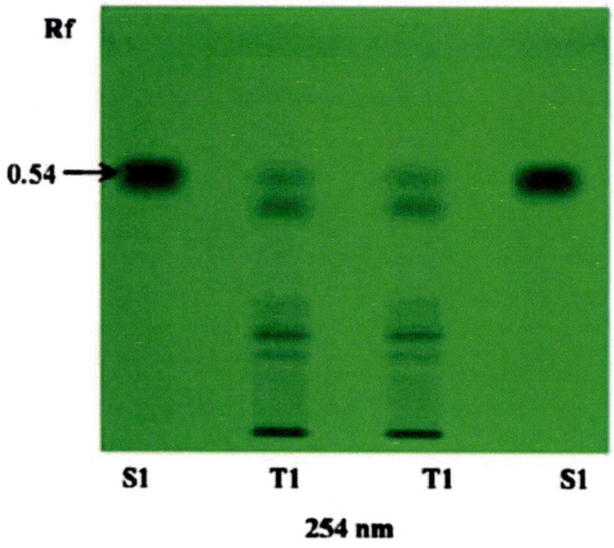

Fig. (63b). HPTLC fingerprint of Rudraksh: S1: Vanillin, T1: Rudraksh seeds extract .

Elettaria cardamomum (L.) Maton (Common Name: Cardamom)

Plant Part: Seeds.

Solvent System: Toluene: ethyl acetate (9.3: 0.7 v/v).

Adsorbent: TLC Silica gel $60F_{254}$ (Merck).

Standard Solution Preparation: 10 mg of eugenol was dissolved in 10 mL of methanol. 20 µL of standard solution was applied on a TLC plate.

Sample Solution Preparation: 500 mg of sample dissolved in 10 mL methanol, sonicated and centrifuged. 15 µL of the solution was applied on TLC plate.

Derivatization: Anisaldehyde sulfuric acid.

Medicinal Value: It is diuretic, anti-inflammatory, helps with digestive problems, ulcers, and prevents cavities (1).

Conclusion: HPTLC fingerprint of Cardamom validates the presence of eugenol, the signature marker at 0.49 Rf, similar to eugenol standard (Fig. **64a**, **b**).

Fig. (64a). Canvas painting of Cardamom illustrated by Patanjali Research Foundation, Haridwar, India.

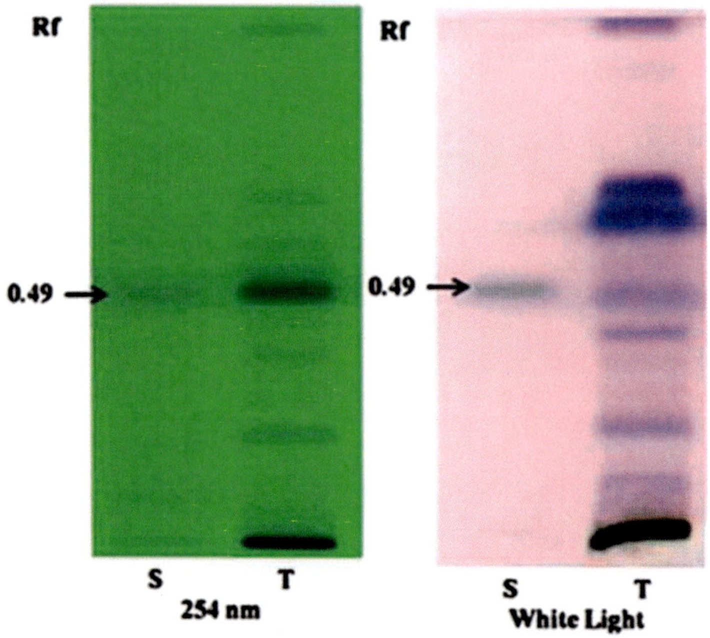

Fig. (64b). HPTLC fingerprint of Cardamom: S: Eugenol, T: Cardamom seeds.

Embelia ribes Burm. F. (Common Name: Viavidang)

Plant Part: Seeds.

Solvent System: Toluene: ethyl acetate: formic acid (7: 3: 1 v/v/v).

Adsorbent: TLC Silica gel $60F_{254}$ (Merck)

Sample Solution Preparation: 1000 mg sample dissolved in 10 mL of methanol, sonicated and centrifuged. 10 µL of the solution was applied on a TLC plate.

Medicinal Value: Useful in throat infection,and indigestion, and treats intestinal worms and skin diseases (3).

Conclusion: HPTLC fingerprint shows fluorescent bands at 366 nm and numbers of bands at 254 nm (Fig. **65a**, **b**).

Fig. (65a). Canvas painting of Viavidang illustrated by Patanjali Research Foundation, Haridwar, India.

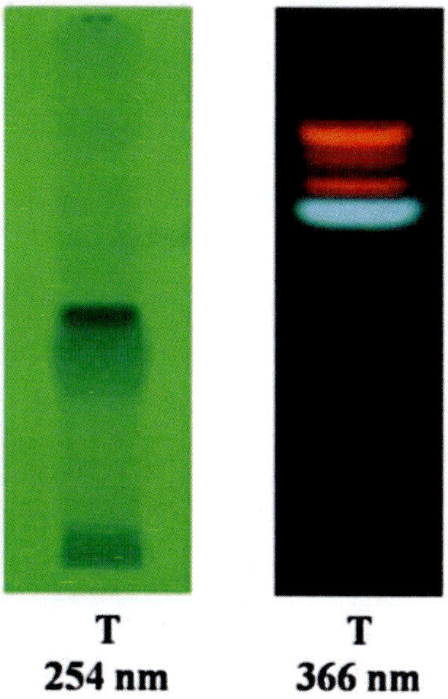

Fig. (65b). HPTLC fingerprint of Viavidang: T: Viavidang seeds.

Foeniculum vulgare Mill (Common Name: Fennel)

Plant Part: Seeds.

Solvent System: Ethyl acetate: toluene: formic acid (16: 4: 1 v/v/v).

Adsorbent: TLC Silica gel $60F_{254}$ (Merck).

Sample Solution Preparation: 1000 mg of the sample dissolved in 10 mL hexane, sonicated and centrifuged. 20 µL of the solution was applied on a TLC plate.

Medicinal Value: Promotes digestive health, aids weight loss and reduces inflammation (3).

Conclusion: HPTLC fingerprint of Fennel shows a major band at 254 nm and 366 nm, which is specific for Fennel in the above mentioned chromatographic conditions (Fig. **66a**, **b**).

Fig. (66a). Canvas painting of Fennel illustrated by Patanjali Research Foundation, Haridwar, India.

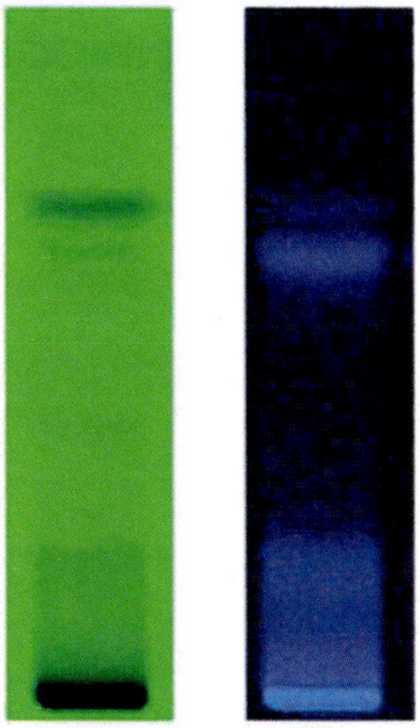

Fig. (66b). HPTLC fingerprint of Fennel: T: Fennel seeds.

Glycyrrhiza glabra L. (Common Name: Licorice)

Plant Part: Stem.

Solvent System: Ethyl acetate: formic acid: acetic acid: water (10: 1.1: 1.1: 2.6 v/v/v/v).

Adsorbent: TLC Silica gel $60F_{254}$ (Merck).

Standard Solution Preparation: 10 mg of Glycyrrhizin was dissolved in 10 mL of methanol. 12 µL of the standard solution was applied on a TLC plate.

Sample Solution Preparation: 500 mg of each sample was dissolved in 10 mL of methanol, sonicated and centrifuged. 1µL of the solution was applied on a TLC plate.

Medicinal Value: Useful in cough, throat pain, skin diseases and respiratory diseases.

Conclusion: HPTLC fingerprint of Licorice confirming the presence of signature bio-active marker glycyrrhizin, at specific Rf 0.22 similar to glycyrrhizin standard (Fig. **67a**, **b**).

Fig. (67a). Canvas painting of Licorice illustrated by Patanjali Research Foundation, Haridwar, India.

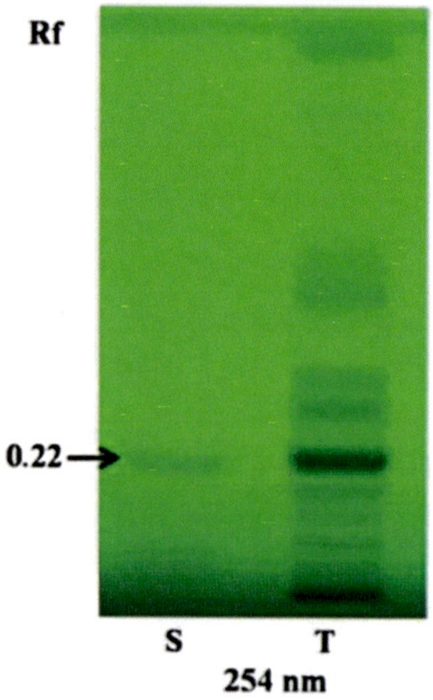

Fig. (67b). HPTLC fingerprint of Licorice: S: Glycyrrhizin, T: Licorice stem.

Habenaria intermedia D.Don (Common Name: Riddhi)

Plant Part: Tuber.

Solvent System: Toluene: ethyl acetate: methanol: formic acid (8: 8: 2: 2 v/v/v/v).

Adsorbent: TLC Silica gel $60F_{254}$ (Merck).

Sample Solution Preparation: 500 mg of the sample dissolved in 10 mL water: methanol (70: 30 v/v), sonicated and centrifuged. 12µL of the solution was applied on a TLC plate.

Derivatization: Anisaldehyde sulfuric acid.

Medicinal Value: Used as an intellect promoter, aphrodisiac, depurative and rejuvenator.

Conclusion: HPTLC fingerprint of Riddhi shows a determined pattern of bands at 254nm and 366 nm and under white light after derivatization (Fig. **68a**, **b**).

Fig. (68a). Canvas painting of Riddhi illustrated by Patanjali Research Foundation, Haridwar, India.

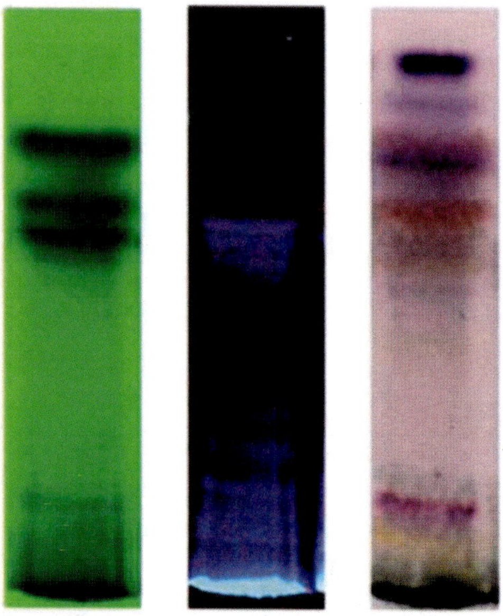

T 254 nm | T 366 nm | T White light

Fig. (68b). HPTLC fingerprint of Riddhi: T: Riddhi tuber.

Hemidesmus indicus (L.) R. Br (Common Name: Anantmool)

Plant Part: Root.

Solvent System: Toluene: ethyl acetate: formic acid (10: 10: 1 v/v/v).

Adsorbent: TLC Silica gel $60F_{254}$ (Merck).

Sample Solution Preparation: 1000 mg sample dissolved in 10 mL of methanol, sonicated and centrifuged. 10 µL of the solution was applied on a TLC plate.

Medicinal Value: Used to treat diarrhoea, gastritis, indigestion and skin-related problems.

Conclusion: HPTLC fingerprint of Anantmool represents three major bands under 254 nm and some fluorescent bands at 366 nm (Fig. **69a**, **b**).

Fig. (69a). Canvas painting of Anantmool illustrated by Patanjali Research Foundation, Haridwar, India.

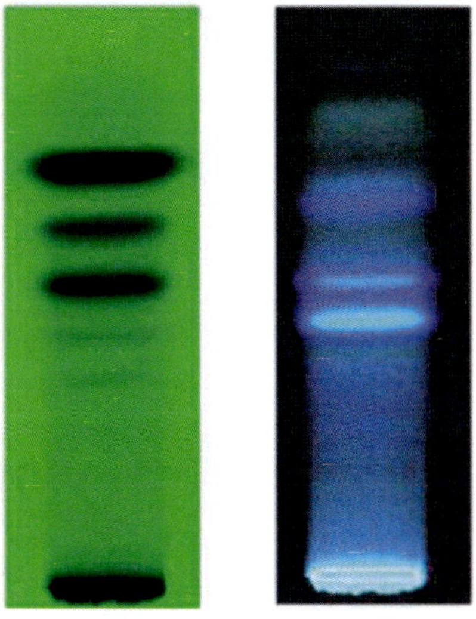

Fig. (69b). HPTLC fingerprint of Anantmool: T: Anantmool root.

Holarrhena pubescens **Wall (Common Name: Indrajao)**

Plant Part: Seeds.

Solvent System: Toluene: ethyl acetate: formic acid (5: 5: 1 v/v/v).

Adsorbent: TLC Silica gel $60F_{254}$ (Merck).

Sample Solution Preparation: 500 mg of the sample dissolved in 10 mL methanol, sonicated and centrifuged. 10 μL of solution was applied on TLC plate.

Medicinal Value: Useful in diarrhoea, fever, indigestion and inflammation (2).

Conclusion: HPTLC fingerprint represents bands of phytochemical components of Indrajao at 254 nm and 366 nm (Fig. **70a**, **b**).

Fig. (70a). Canvas painting of Indrajao illustrated by Patanjali Research Foundation, Haridwar, India.

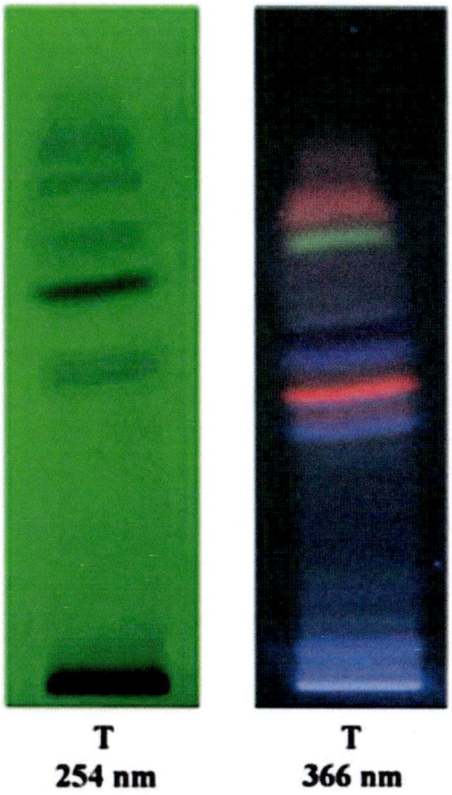

Fig. (70b). HPTLC fingerprint of Indrajao: T: Indrajao seeds.

Holarrhena pubescens Wall (Common Name: Kutaj)

Plant Part: Bark.

Solvent System: Toluene: ethyl acetate: formic acid (5: 5: 1 v/v/v).

Adsorbent: TLC Silica gel $60F_{254}$ (Merck).

Sample Solution Preparation: 500 mg of the sample dissolved in 10 mL methanol, sonicated and centrifuged. 10 µL of the solution was applied on TLC plate.

Medicinal Value: Useful in diarrhoea, fever, indigestion and inflammation (2).

Conclusion: HPTLC fingerprint represents specific bands of phytochemical molecules of Kutaj at 254 nm and 366 nm (Fig. **71a**, **b**).

Fig. (71a). Canvas painting of Kutaj illustrated by Patanjali Research Foundation, Haridwar, India.

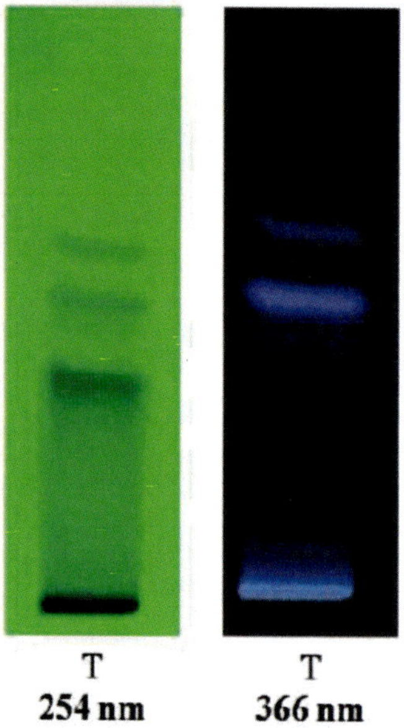

Fig. (71b). HPTLC fingerprint of Kutaj: T: Kutaj bark.

***Lagenaria siceraria* (Molina) Standl (Common Name: Bottle Gourd)**

Plant Part: Fruit.

Solvent System: Butanol: acetic acid: water (6.5: 1.5: 2 v/v/v).

Adsorbent: TLC Silica gel $60F_{254}$ (Merck).

Sample Solution Preparation: 500 mg of the sample dissolved in 10 mL water, sonicated and centrifuged. 10 µL of the solution was applied on a TLC plate.

Derivatization: Anisaldehyde sulfuric acid.

Medicinal Value: Used to treat acidity, indigestion and ulcer.

Conclusion: HPTLC fingerprint of bottle gourd shows a specific banding pattern at 254 nm, at 366 nm and under white light after derivatization (Fig. **72a**, **b**).

Fig. (72a). Canvas painting of Bottle gourd illustrated by Patanjali Research Foundation, Haridwar, India.

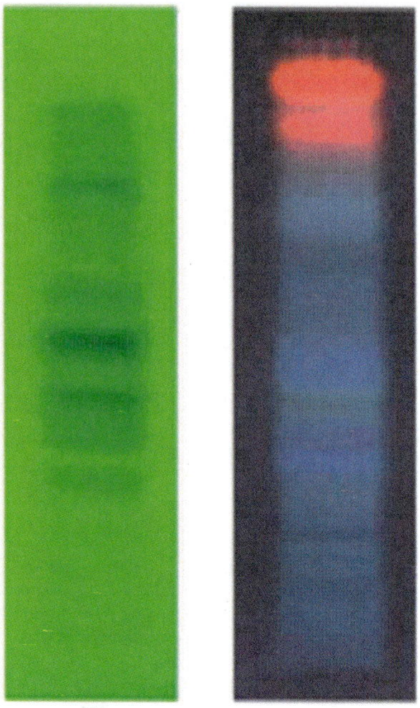

Fig. (72b). HPTLC fingerprint of Bottle gourd: T: Bottle gourd fruit.

Lavandula stoechas L. (Common Name: Ustekhaddus)

Plant Part: Flowers.

Solvent System: Ethyl acetate: toluene: formic acid (16: 4: 1 v/v/v).

Adsorbent: TLC Silica gel $60F_{254}$ (Merck).

Sample Solution Preparation: 1000 mg sample dissolved in 10 mL of methanol, sonicated and centrifuged. 20 μL of the solution was applied on a TLC plate.

Medicinal Value: Plant has anti- inflammatory, antioxidant, antispasmodic, antimicrobial and antifungal activities (1).

Conclusion: HPTLC fingerprint of Ustekhaddus is captured under UV display, showing specific major three bands at 254 nm and fluorescent bands at 366 nm (Fig. **73a, b**).

Fig. (73a). Canvas painting of Ustekhaddus illustrated by Patanjali Research Foundation, Haridwar, India.

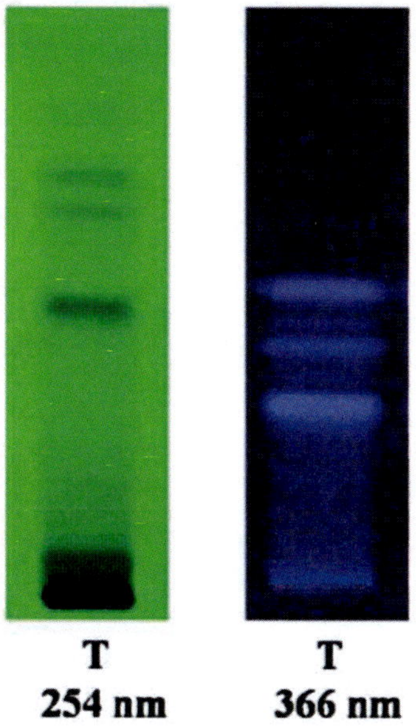

Fig. (73b). HPTLC fingerprint of Ustekhaddus: T: Ustekhaddus flowers.

Lawsonia inermis L. (Common Name: Henna)

Plant Part: Leaves.

Solvent System: Ethyl acetate: toluene: formic acid (5: 5: 1 v/v/v).

Adsorbent: TLC Silica gel $60F_{254}$ (Merck).

Sample Solution Preparation: 500 mg of sample dissolved in 10 mL methanol, sonicated and centrifuged. 8 µL of the solution was applied on TLC plate.

Medicinal Value: Used to treat dysuria, inflammation, mouth ulcer, boils, skin disease and hair related issues (3).

Conclusion: Chromatographic fingerprint represents phytochemical components of Henna, at 254 nm and 366 nm. Major red colored fluorescent bands appeared at 366 nm (Fig. **74a**, **b**).

Fig. (74a). Canvas painting of Henna illustrated by Patanjali Research Foundation, Haridwar, India.

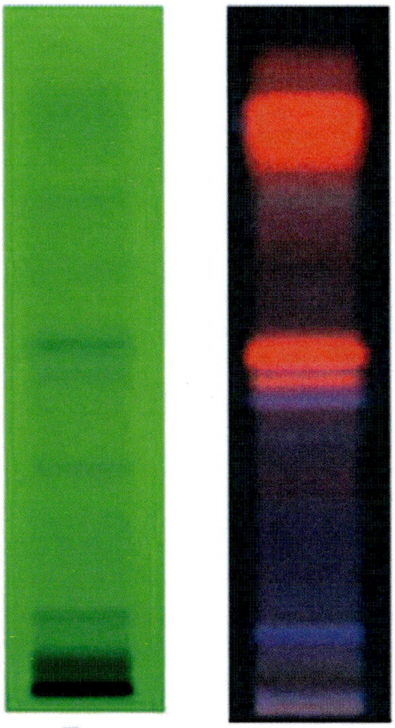

Fig. (74b). HPTLC fingerprint of Henna: T: Henna leaves.

Leucas cephalotes (Roth) Spreng (Common Name: Dronpushpi)

Plant Part: Leaves.

Solvent System: Toluene: ethyl acetate (9.3: 0.7 v/v).

Adsorbent: TLC Silica gel $60F_{254}$ (Merck).

Sample Solution Preparation: 500 mg of the sample dissolved in 10 mL methanol, sonicated and centrifuged. 15 µL of the solution was applied on a TLC plate.

Medicinal Value: Plant has antimicrobial, insecticidal, antipyretic and anti-inflammatory properties (2).

Conclusion: HPTLC fingerprint of Dronpushpi shows bands of phytochemicals at 254 nm and red fluorescent bands at 366 nm (Fig. **75a**, **b**).

Fig. (75a). Canvas painting of Dronpushpi illustrated by Patanjali Research Foundation, Haridwar, India.

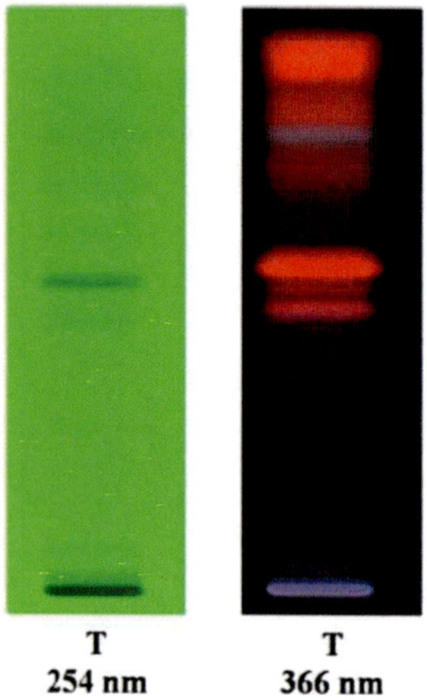

Fig. (75b). HPTLC fingerprint of Dronpushpi: T: Dronpushpi leaves.

Lilium polyphyllum D.Don (Common Name: Ksheer kakoli)

Plant Part: Tuber.

Solvent System: Ethyl acetate: toluene: formic acid: methanol (8: 8: 2: 2 v/v/v/v).

Adsorbent: TLC Silica gel $60F_{254}$ (Merck).

Sample Solution Preparation: 500 mg of each sample dissolved in 10 mL of water: methanol (70: 30 v/v), sonicated and centrifuged. 10μL of the solution was applied on a TLC plate.

Medicinal Use: The plant is useful in treating respiratory disorders like bronchial asthma and cough. It is also a rejuvenative and immunity booster.

Conclusion: In HPTLC fingerprint of Ksheer kakoli, major fluorescent band, visualized at 366 nm and after derivatization, purple bands of phytochemicals were observed under white light (Fig. **76a**, **b**).

Fig. (76a). Canvas painting of Ksheer kakoli illustrated by Patanjali Research Foundation, Haridwar, India.

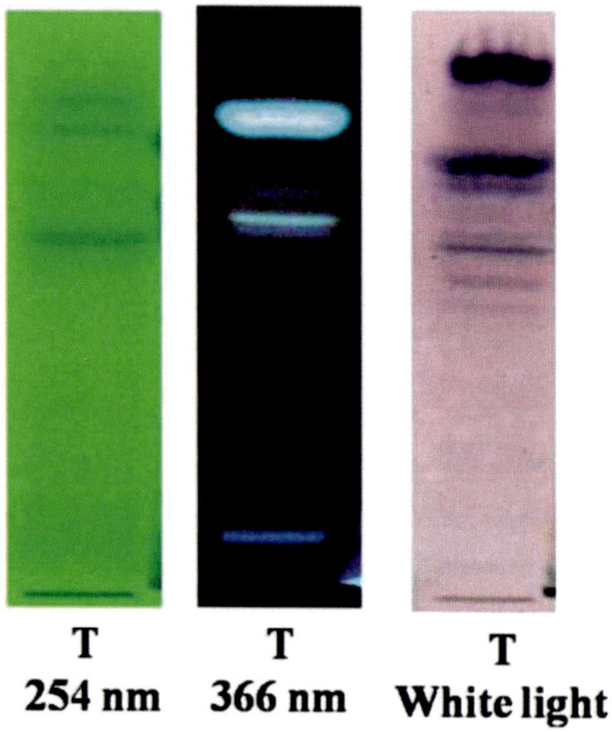

Fig. (76b). HPTLC fingerprint of Ksheer kakoli: T: Ksheer kakoli tuber.

Mentha piperita L. (Common Name: Mint)

Plant Part: Leaves.

Solvent System: Hexane: ethyl acetate (8: 2 v/v).

Adsorbent: TLC Silica gel $60F_{254}$ (Merck).

Sample Solution Preparation: 500 mg of each sample dissolved in 10 mL of water: methanol (70: 30 v/v), sonicated and centrifuged. 10μL of the solution was applied on a TLC plate.

Medicinal Use: Useful in treating common cold, and gastro-intestinal problems, easing symptoms of nausea, and improving oral hygiene and dental health (2).

Conclusion: HPTLC fingerprint of Mint shows a specific banding pattern at 254 nm and 366 nm (Fig. **77a**, **b**).

Fig. (77a). Drawing of Mint illustrated by Patanjali Research Foundation, Haridwar, India.

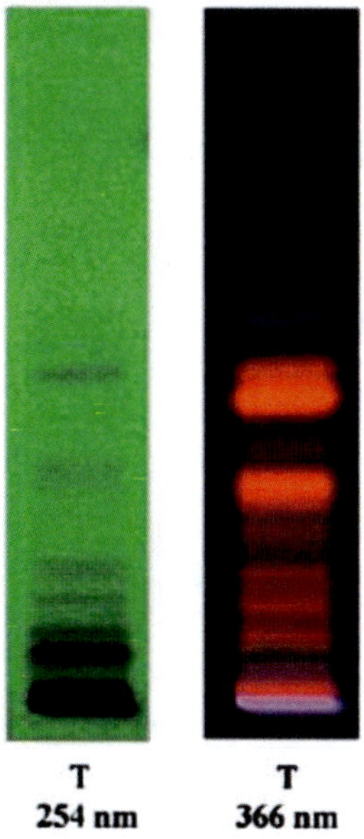

Fig. (77b). HPTLC fingerprint of Mint: T: Mint leaves.

Momordica charantia L. (Common Name: Bitter Gourd)

Plant Part: Fruit.

Solvent System: Ethyl acetate: toluene: formic acid (5: 5: 1 v/v/v).

Adsorbent: TLC Silica gel $60F_{254}$ (Merck).

Sample Solution Preparation: 1000 mg of the sample dissolved in 10 mL methanol, sonicated and centrifuged. 8μL of the solution was applied on a TLC plate.

Medicinal Value: It is anthelminthic, diuretic, and anti-diabetic; used in hair and skin-related diseases. (1).

Conclusion: Fingerprint of Bitter gourd showing specific pattern of bands at 254 nm and 366 nm (Fig. **78a, b**).

Fig. (78a). Canvas painting of Bitter gourd illustrated by Patanjali Research Foundation, Haridwar, India.

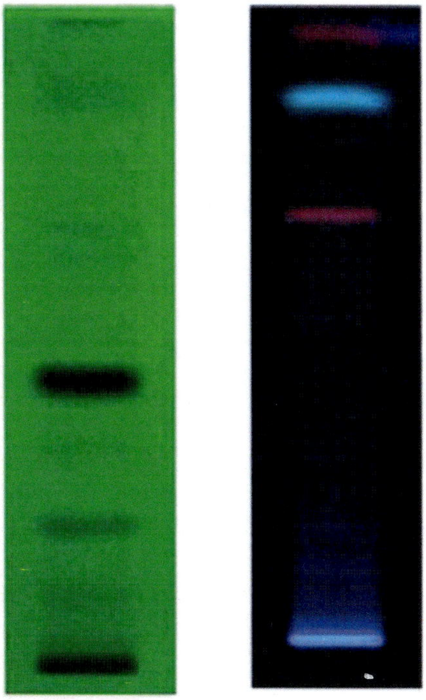

Fig. (78b). HPTLC fingerprint of Bitter gourd: T: Bitter gourd fruit.

Moringa oleifera Lam (Common Name: Sahjan)

Plant Part: Leaves.

Solvent System: Toluene: ethyl acetate: formic acid (5: 5: 1 v/v/v).

Adsorbent: TLC Silica gel $60F_{254}$ (Merck).

Sample Solution Preparation: 1000 mg of the sample dissolved in 10 mL methanol, sonicated and centrifuged. 10 μL of the solution was applied on a TLC plate.

Medicinal Use: The plant is used as cardiac and circulatory stimulant, possess antitumor, antipyretic, anti-inflammatory, antiulcer, antispasmodic, diuretic, antioxidant, antibacterial and antifungal activities (3).

Conclusion: HPTLC fingerprint represents specific bands of phytochemical molecules of Sahjan at 254 nm and 366 nm (Fig. **79a**, **b**).

Fig. (79a). Canvas painting of Sahjan illustrated by Patanjali Research Foundation, Haridwar, India.

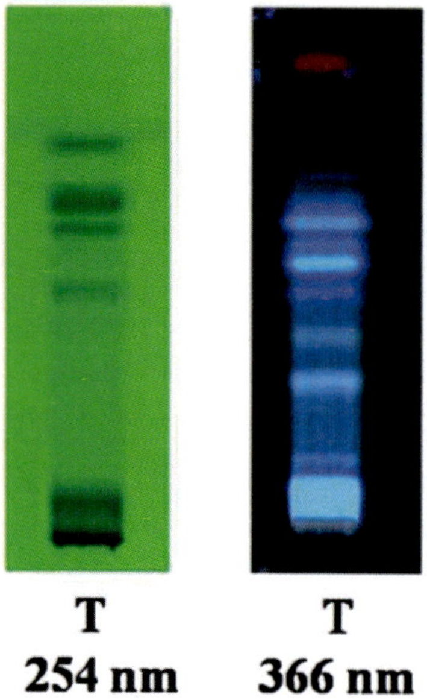

Fig. (79b). HPTLC fingerprint of Sahjan: T: Sahjan leaves.

Murraya paniculata (L.) Jack (Common Name: Gandhani)

Plant Part: Leaves.

Solvent System: Ethyl acetate: toluene: formic acid (5: 5: 0.5 v/v/v).

Adsorbent: TLC Silica gel $60F_{254}$ (Merck).

Sample Solution Preparation: 1000 mg of the sample dissolved in 10 mL methanol, sonicated and centrifuged. 3 µL of the solution was applied on a TLC plate.

Medicinal Value: Used to treat fever, cough sore throat, dysentery and diarrhoea.

Conclusion: Fingerprint of Gandhani showing specific patterns of bands at 254 nm and 366 nm (Fig. **80a**, **b**).

Fig. (80a). Canvas painting of Gandhani illustrated by Patanjali Research Foundation, Haridwar, India.

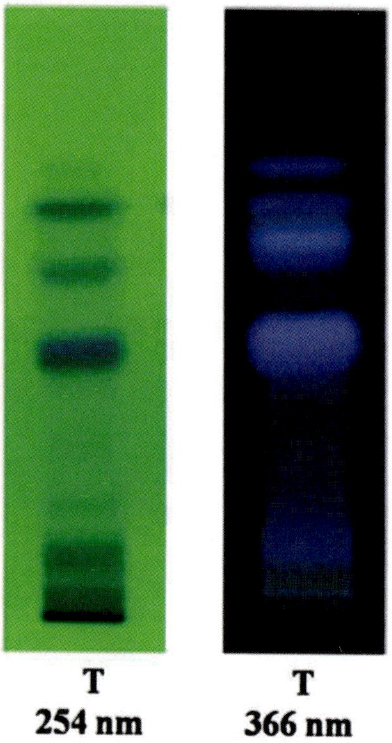

Fig. (80b). HPTLC fingerprint of Gandhani: T: Gandhani leaves.

Myristica fragrans Houtt (Common Name: Mace)

Plant Part: Aril.

Solvent System: Toluene: ethyl acetate formic acid (8: 2: 0.3 v/v/v).

Adsorbent: TLC Silica gel $60F_{254}$ (Merck).

Sample Solution Preparation: 500 mg sample dissolved in 10 mL of methanol, sonicated and centrifuged. 10 µL of the solution was applied on a TLC plate.

Medicinal Value: Useful in clearing throat, cough, asthma, throat pain, irritable bowel syndrome and diarrhoea (2).

Conclusion: HPTLC Fingerprint of Mace showing a specific pattern of bands at 254 nm (Fig. **81a**, **b**).

Fig. (81a). Canvas painting of Mace illustrated by Patanjali Research Foundation, Haridwar, India.

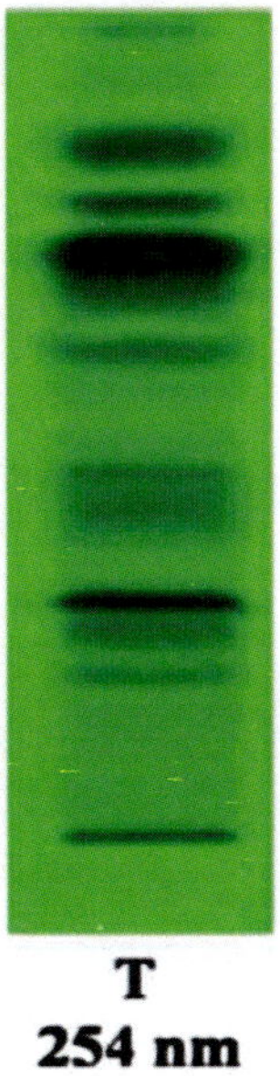

Fig. (81b). HPTLC fingerprint of Mace: T: Mace aril.

Myristica fragrans Houtt (Common Name: Jaiphal)

Plant Part: Fruit.

Solvent System: Toluene: ethyl acetate formic acid (7: 3: 1 v/v/v).

Adsorbent: TLC Silica gel 60F_{254} (Merck).

Sample Solution Preparation: 1000 mg sample dissolved in 10 mL of methanol, sonicated and centrifuged. 10 μL of the solution was applied on a TLC plate.

Medicinal Value: Useful in clearing throat, cough, asthma, throat pain, irritable bowel syndrome and diarrhoea (2).

Conclusion: HPTLC fingerprint of Nutmeg represents bands of phytochemicals at 254 nm, fluorescent bands at 366 nm and blue, bluish violet bands under white light after derivatization (Fig. **82a**, **b**).

Fig. (82a). Canvas painting of Jaiphal illustrated by Patanjali Research Foundation, Haridwar, India.

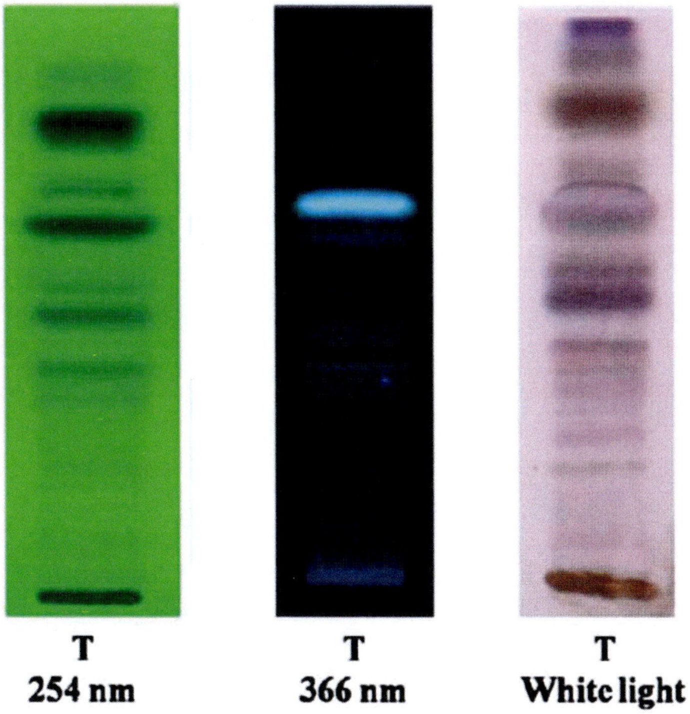

Fig. (82b). HPTLC fingerprint of Nutmeg: T: Nutmeg fruit.

Nardostachys jatamansi (D.Don) DC (Common Name: Jatamansi)

Plant Part: Roots.

Solvent System: Toluene: ethyl acetate formic acid (8: 2: 0.3 v/v/v).

Adsorbent: TLC Silica gel $60F_{254}$ (Merck).

Sample Solution Preparation: 1000 mg sample dissolved in 10 mL methanol, sonicated and centrifuged. 10 µL of the solution was applied on a TLC plate.

Medicinal Value: Useful in inducing sound sleep, improving intelligence and reducing anxiety (2).

Conclusion: HPTLC photograph at 254 nm and 366 nm represents a specific band pattern for Jatamansi in given chromatographic conditions (Fig. **83a**, **b**).

Fig. (83a). Canvas painting of Jatamansi illustrated by Patanjali Research Foundation, Haridwar, India.

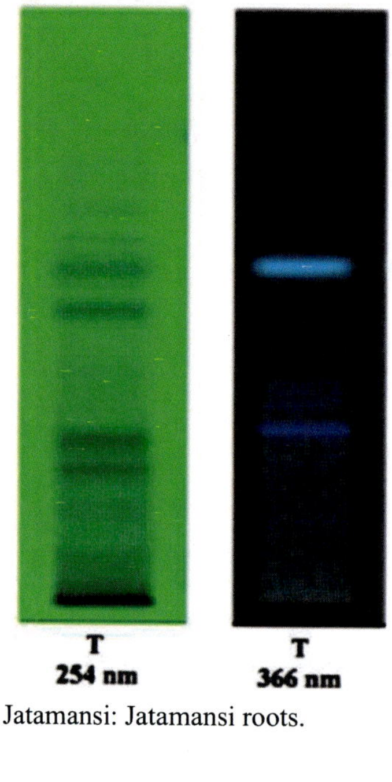

Fig. (83b). HPTLC fingerprint of Jatamansi: Jatamansi roots.

Nymphaea nouchali Burm. F. (Common Name: Blue Lotus)

Plant Part: Flower.

Solvent System: Toluene: ethyl acetate formic acid (7: 3: 1 v/v/v).

Adsorbent: TLC Silica gel 60F_{254} (Merck)

Sample Solution Preparation: 1000 mg sample dissolved in 10 mL methanol, sonicated and centrifuged. 10 µL of the solution was applied on a TLC plate.

Medicinal Value: Useful in bleeding disorders, body ache, and skin boils and provides relief from burning sensations (1).

Conclusion: HPTLC fingerprint of blue lotus shows spots of phytochemicals under UV and white light. Fluorescent bands observed at 366 nm and blue and purple bands under white light after derivatization (Fig. **84a**, **b**).

Fig. (84a). Canvas painting of Blue lotus illustrated by Patanjali Research Foundation, Haridwar, India.

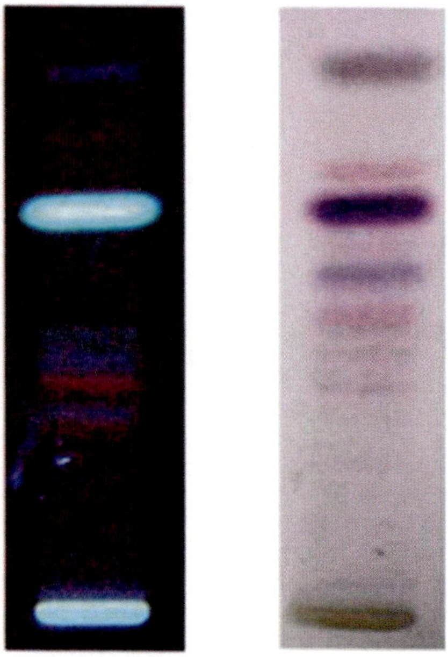

Fig. (84b). HPTLC fingerprint of Blue lotus: T: Blue lotus flower.

Ocimum basilicum L. (Common Name: Basil)

Plant Part: Leaves.

Solvent System: Ethyl acetate: toluene: formic acid (10: 9: 1 v/v/v).

Adsorbent: TLC Silica gel $60F_{254}$ (Merck).

Sample Solution Preparation: 1 g of sample dissolved in 10 mL methanol: water (60: 40 v/v), sonicated and centrifuged. 8 μL solution was applied on the TLC plate.

Medicinal Value: Plant has antibacterial, antifungal and insecticidal activities (2).

Conclusion: HPTLC fingerprint shows bands of phytochemical components at 254 nm, 366 nm and under white light after derivatization (Fig. **85a**, **b**).

Fig. (85a). Drawing of Basil illustrated by Patanjali Research Foundation, Haridwar, India.

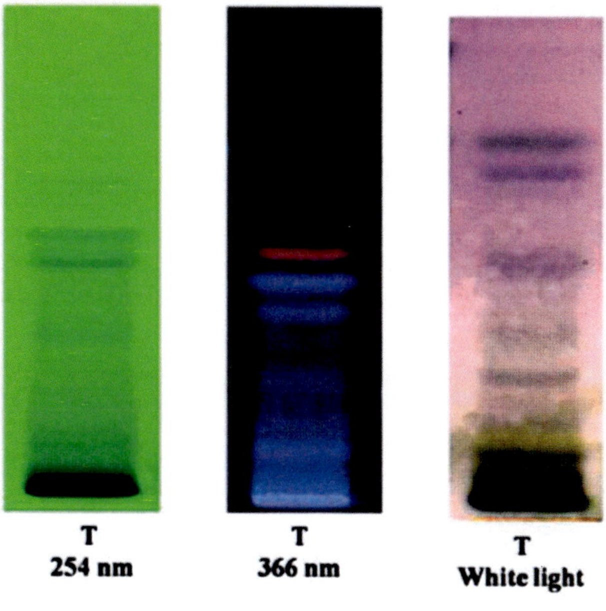

Fig. (85b). HPTLC fingerprint of Basil: T: Basil.

Ocimum tenuiflorum L. (Common Name: Basil)

Plant Part: Leaves.

Solvent System: Ethyl acetate: toluene: formic acid (10: 9: 1 v/v/v).

Adsorbent: TLC Silica gel $60F_{254}$ (Merck).

Standard Solution Preparation: 10 mg of each standard was dissolved in 10 mL of methanol. 5 µL of each standard solution was applied on a TLC plate.

Sample Solution Preparation: 1000 mg of the sample dissolved in 10 mL methanol: water (60: 40 v/v), sonicated and centrifuged. 8 µL solution was applied on a TLC plate.

Medicinal Value: Used to treat fever, cold, cough, oral health and respiratory disease (2).

Conclusion: HPTLC fingerprint shows the presence of bio-active markers gallic acid, rosmarinic acid, apigenin, caffeic acid and methyl caffate in Basil. Rf values of bands in sample matches with standards (Fig. **86a**, **b**).

Fig. (86a). Canvas painting of Basil illustrated by Patanjali Research Foundation, Haridwar, India.

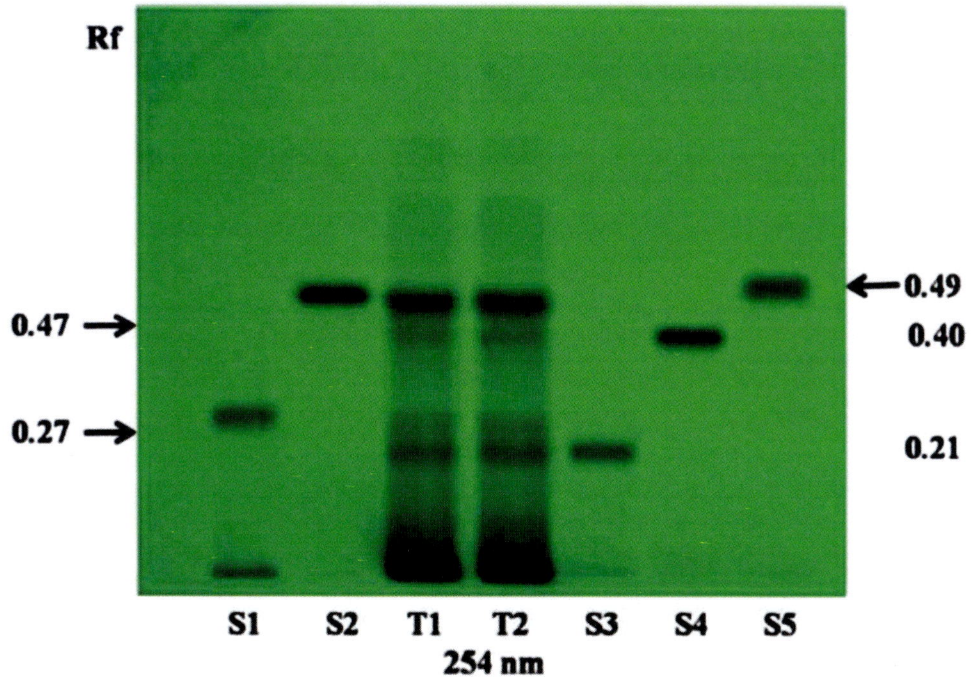

Fig. (86b). HPTLC fingerprint of Basil: S1: Gallic acid, S2: Rosmarinic acid, T1 and T2: Basil leaves, S3: Apigenin, S4: Caffeic acid, S5: Methyl caffate.

Operculina turpethum **(L.) Silva Manso (Common Name: Nisoth)**

Plant Part: Stem.

Solvent System: Toluene: ethyl acetate: formic acid (10: 10: 1 v/v/v).

Adsorbent: TLC Silica gel $60F_{254}$ (Merck).

Sample Solution Preparation: 1000 mg of sample dissolved in 10 mL methanol, sonicated and centrifuged. 10 µL of solution was applied on a TLC plate.

Derivatizing Reagent: Anisaldehyde sulfuric acid.

Medicinal Value: Useful to treat cough, bronchitis, arthritis and constipation (2).

Conclusion: HPTLC fingerprint of Nisoth represents bands of phytochemicals at 254 nm, fluorescent bands at 366 nm before and after derivatization (Fig. **87a**, **b**).

Fig. (87a). Canvas painting of Nisoth illustrated by Patanjali Research Foundation, Haridwar, India.

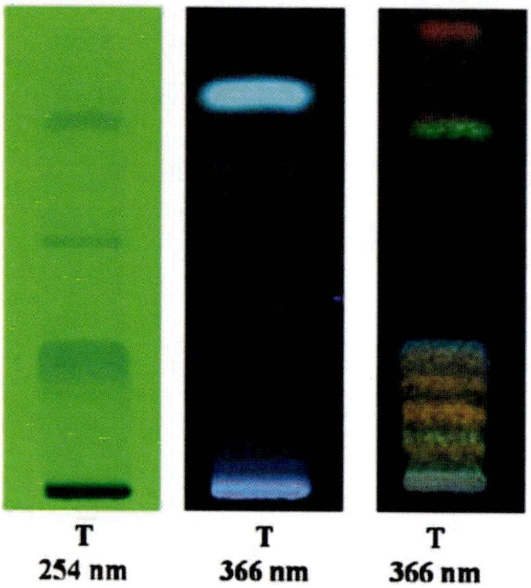

Fig. (87b). HPTLC fingerprint of Nisoth: T: Nisoth stem.

Phyllanthus emblica L. (Common Name: Indian gooseberry)

Plant Part: Fruit.

Solvent System: Ethyl acetate: toluene: formic acid (5: 2: 1.5 v/v/v).

Adsorbent: TLC Silica gel $60F_{254}$ (Merck).

Standard Solution Preparation: 10 mg of gallic acid was dissolved in 10 mL of methanol. 20 µL of the standard solution was applied on a TLC plate.

Sample Solution Preparation: 1000 mg of the sample dissolved in 10 mL methanol sonicated and centrifuged. 6 µL of the solution was applied on a TLC plate.

Medicinal Value: Improves immunity, reduces skin blemishes, improves eyesight, acts as blood purifier and also prevents graying of hairs (1).

Conclusion: The bio-active marker gallic acid is confirmed in HPTLC fingerprint. It shows the band of gallic acid at Rf 0.61 with respect to standard (Fig. **88a**, **b**).

Fig. (88a). Canvas painting of Indian gooseberry illustrated by Patanjali Research Foundation, Haridwar, India.

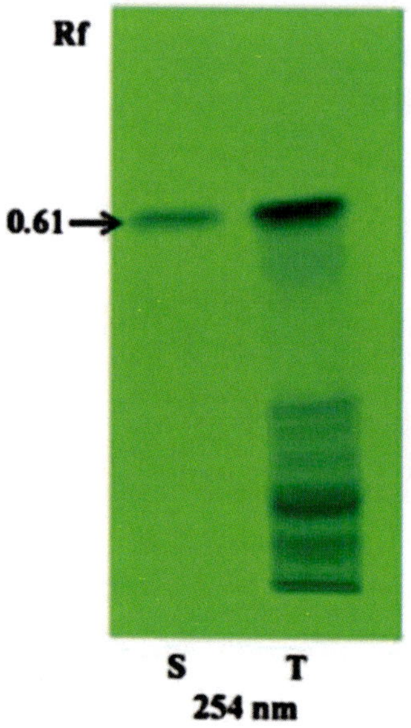

Fig. (88b). HPTLC fingerprint of Indian gooseberry: S: Gallic acid, T: Indian gooseberry fruit.

Picrorhiza kurroa Royle ex Benth (Common Name: Kutki)

Plant Part: Root.

Solvent System: Ethyl acetate: methanol: water (8.2: 1.0: 0.8 v/v/v)

Adsorbent: TLC Silica gel $60F_{254}$ (Merck).

Standard Solution Preparation: 1 mg of picroside I was dissolved in 10 mL of methanol. 10 µL of the standard solution was applied on a TLC plate.

Sample Solution Preparation: 500 mg of each sample dissolved in 10 mL of their respective solvents, sonicated and centrifuged. 6 µL of the solution was applied on a TLC plate.

Medicinal Use: Useful in asthma, cough, cold and indigestion (1).

Conclusion: HPTLC fingerprint of Kutki confirming the presence of signature bio-active marker picroside I, at Rf 0.45 as similar to picroside I standard (Fig. **89a, b**).

Fig. (89a). Canvas painting of Kutki illustrated by Patanjali Research Foundation, Haridwar, India.

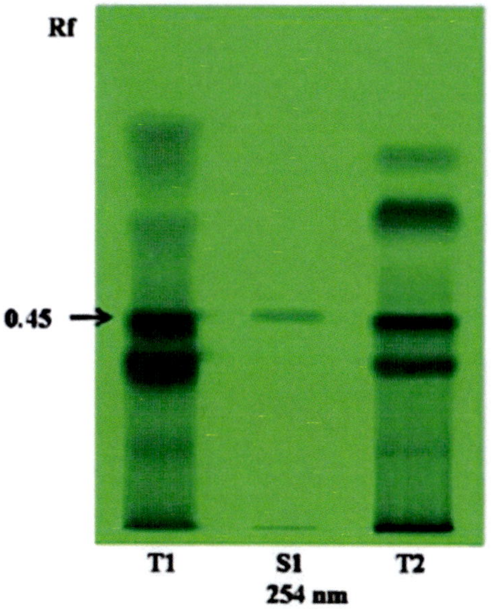

Fig. (89b). HPTLC fingerprint of Kutki: T1: Kutki root ethanolic extract, S1: Picroside I, T2: Kutki root water extract.

Piper longum L. (Common Name: Chhoti Pippali)

Plant Part: Fruit.

Solvent System: Toluene: ethyl acetate: formic acid (5: 5: 0.5 v/v/v).

Adsorbent: TLC Silica gel $60F_{254}$ (Merck).

Standard Solution Preparation: 10 mg of piperine was dissolved in 10 mL of methanol. 2 μL of standard solution was applied on a TLC plate.

Sample Solution Preparation: 500 mg of the sample dissolved in 10 mL methanol, sonicated and centrifuged. 10 μL of the solution was applied on a TLC plate.

Medicinal Value: Useful in cough, asthma, fever, bronchitis and sore throat (2).

Conclusion: Rf 0.52 of the sample confirmed the presence of piperine in Chhoti pippali by HPTLC fingerprint (Fig. **90a**, **b**).

Fig. (90a). Canvas painting of Chhoti pippali illustrated by Patanjali Research Foundation, Haridwar, India.

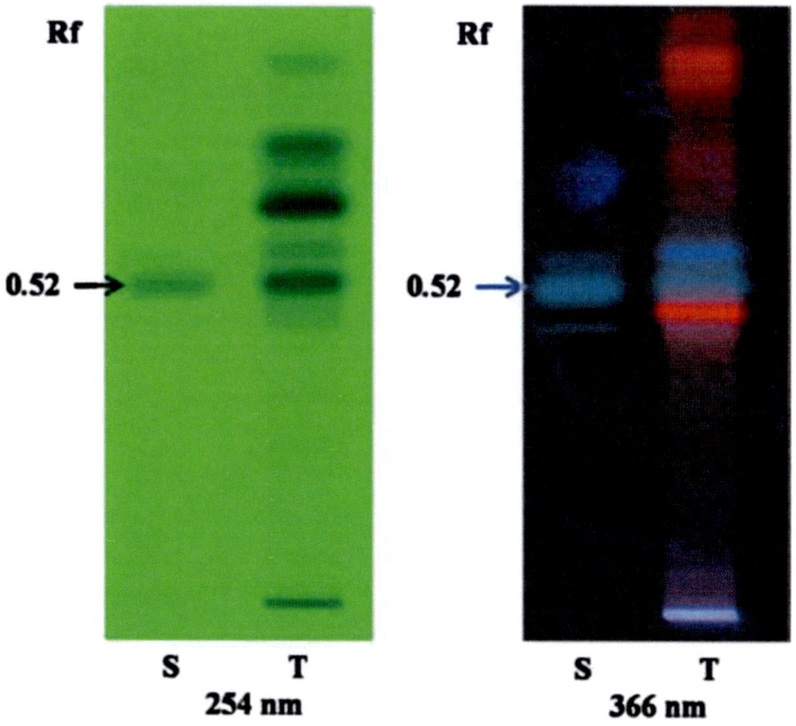

Fig. (90b). HPTLC fingerprint of Chhoti pippali: S: Piperine,T: Chhoti pippali fruit.

Piper nigrum L. (Common Name: Black pepper)

Plant Part: Seeds.

Solvent System: Toluene: ethyl acetate: formic acid (5: 5: 0.5 v/v/v).

Adsorbent: TLC Silica gel $60F_{254}$ (Merck).

Standard Solution Preparation: 10 mg of piperine was dissolved in 10 mL of methanol. 2 µL of the standard solution was applied on a TLC plate.

Sample Solution Preparation: 500 mg of the sample dissolved in 10 mL methanol, sonicated and centrifuged. 10 µL of the solution was applied on TLC plate.

Medicinal Value: Useful in cold cough and fever and respiratory disease (3).

Conclusion: HPTLC fingerprint of Black pepper shows the presence of Piperine. Rf 0.52 confirms the band of piperine in black pepper, as similar to standard (Fig. **91a, b**).

Fig. (91a). Canvas painting of Black pepper illustrated by Patanjali Research Foundation, Haridwar, India.

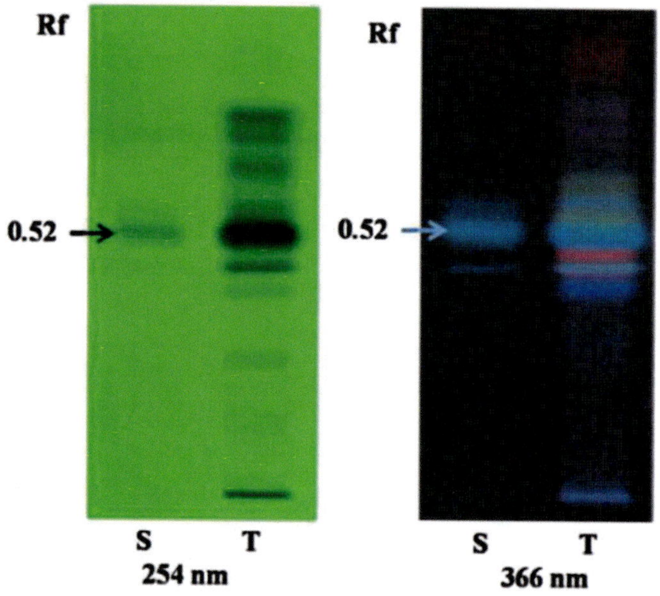

Fig. (91b). HPTLC fingerprint of Black Pepper: S: Piperine, T: Black pepper seeds.

Piper retrofractum Vahl (Common Name: Badi Pippali)

Plant Part: Fruit.

Solvent System: Toluene: ethyl acetate: formic acid (5: 5: 0.5 v/v/v).

Adsorbent: TLC Silica gel $60F_{254}$ (Merck).

Standard Solution Preparation: 10 mg of piperine was dissolved in 10 mL of methanol. 2 µL of the standard solution was applied on TLC plate.

Sample Solution Preparation: 500 mg of the sample dissolved in 10 mL methanol, sonicated and centrifuged. 10 µL of the solution was applied on TLC plate.

Medicinal Value: Useful in cold cough and fever and respiratory disease (2).

Conclusion: Photograph of TLC attesting the presence of marker piperine in Badi pippali. Rf value 0.52 conforming the presence of Piperine (Fig. **92a**, **b**).

■ ■

Fig. (92a). Canvas painting of Badi pippali illustrated by Patanjali Research Foundation, Haridwar, India.

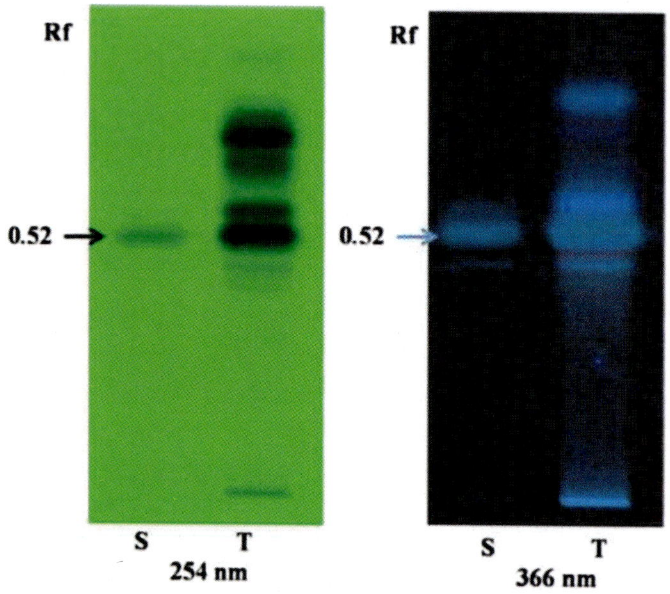

Fig. (92b). HPTLC fingerprint of Badi pippali: S: Piperine, T: Badi Pippali fruit.

Pistacia chinensis subsp. Integerrima (J.L.Stewart) Rech.f (Common Name: Kakra singhi)

Plant Part: Galls.

Solvent System: Toluene: ethyl acetate: formic acid (6: 4: 1 v/v/v).

Adsorbent: TLC Silica gel $60F_{254}$ (Merck).

Standard Solution Preparation: 10 mg of methyl gallate was dissolved in 10 mL of methanol. 10 µL of the standard solution was applied on TLC plate.

Sample Solution Preparation: 1000 mg of the sample dissolved in 10 mL of methanol: water (50: 50 v/v), sonicated and centrifuged. 3 µL of solution was applied on TLC plate.

Medicinal Value: Used to treat cough, tuberculosis, bronchial asthma and diarrhoea (1).

Conclusion: HPTLC fingerprint confirms the presence of methyl gallate in sample as the fingerprint of Kakra singhi showing a band of standard at 0.51 Rf (Fig. **93a**, **b**).

Fig. (93a). Canvas painting of Kakra singhi illustrated by Patanjali Research Foundation, Haridwar, India.

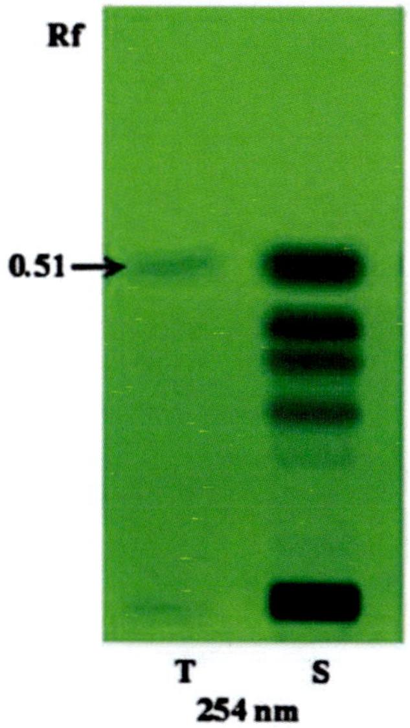

Fig. (93b). HPTLC fingerprint of Kakra singhi: T: Methyl gallate, S: Kakra singhi galls.

Putranjiva roxburghii Wall (Common Name: Putrajeevak)

Plant Part: Seeds.

Solvent System: Chloroform: methanol (9.5: 0.5 v/v).

Adsorbent: TLC Silica gel $60F_{254}$ (Merck).

Standard Solution Preparation: 10 mg of beta-sitosterol was dissolved in 10 mL of methanol. 12 μL of the standard solution was applied on TLC plate.

Sample Solution Preparation: 1000 mg of the sample dissolved in 10 mL methanol, sonicated and centrifuged. 8 μL of the solution was applied on a TLC plate.

Derivatization: Anisaldehyde sulfuric acid

Medicinal Value: It is fertility booster, aphrodisiac, spermatogenic, and anti-inflammatory (2).

Conclusion: Rf 0.58 confirms the presence of beta-sitosterol in Putrajeeva, as evident by HPTLC fingerprint (Fig. **94a**, **b**).

Fig. (94a). Canvas painting of Putrajeevak illustrated by Patanjali Research Foundation, Haridwar, India.

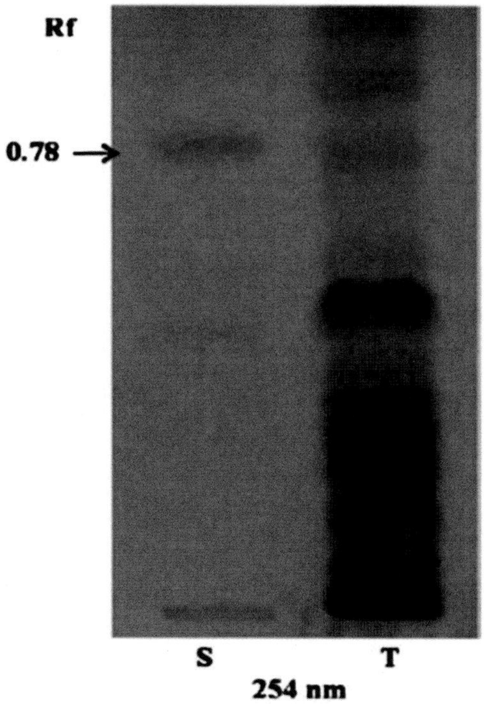

Fig. (94b). HPTLC fingerprint of Putrajeevak: S: Beta-sitosterol, T: Putrajeevak seeds.

Raphanus raphanistrum L. subsp. Sativus (L.) Domin (Common Name: Radish)

Plant Part: Root.

Solvent System: Toluene: ethyl acetate: formic acid (5: 5: 1 v/v/v).

Adsorbent: TLC Silica gel $60F_{254}$ (Merck).

Sample Solution Preparation: 500 mg of the sample dissolved in 10 mL of water: methanol (70: 30 v/v) sonicated and centrifuged. 10 µl of the solution was applied on a TLC plate.

Derivatizing Reagent: Anisaldehyde sulfuric acid.

Medicinal Value: Useful to treat acne, fungal and bacterial infections, and prevent gastrointestinal diseases (3).

Conclusion: HPTLC fingerprint of Radish represents the bands of phytochemicals at 254 nm, 366 nm and under white light after derivatization (Fig. **95a**, **b**).

Fig. (95a). Canvas painting of Radish illustrated by Patanjali Research Foundation, Haridwar, India.

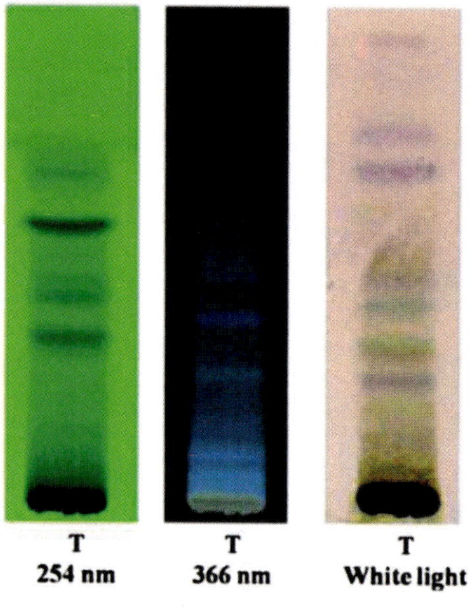

Fig. (95b). HPTLC fingerprint of Radish: T: Radish root.

Rheum webbianum Royle (Common Name: Revandchini)

Plant Part: Stem.

Solvent System: Ethyl acetate: toluene: formic acid (5: 5: 0.5 v/v/v).

Adsorbent: TLC Silica gel $60F_{254}$ (Merck).

Sample Solution Preparation: 500 mg of the sample dissolved in 10 mL methanol, sonicated and centrifuged. 8 µL of the solution was applied on a TLC plate.

Medicinal Value: Used to treat constipation, diarrhoea, dysentery and liver disorders (3).

Conclusion: HPTLC fingerprint of Revandchini represents the specific banding pattern at 254 nm and 366 nm (Fig. **96a**, **b**).

Fig. (96a). Canvas painting of Revandchini illustrated by Patanjali Research Foundation, Haridwar, India.

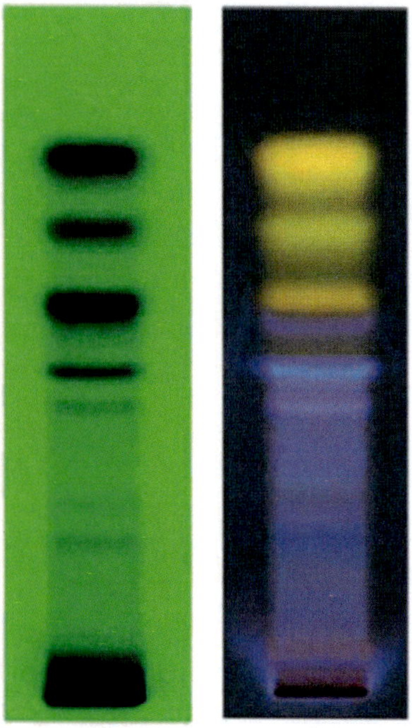

T
254 nm

T
366 nm

Fig. (96b). HPTLC fingerprint of Revandchini: T: Revandchini stem.

Rosa indica L. (Common Name: Rose)

Plant Part: Flower.

Solvent System: Toluene: ethyl acetate formic acid (10: 10: 2 v/v/v).

Adsorbent: TLC Silica gel $60F_{254}$ (Merck).

Sample Solution Preparation: 1000 mg sample dissolved in 10 mL methanol, sonicated and centrifuged. 10 μL of the solution was applied on a TLC plate.

Medicinal Value: Strengthen the digestive system, and useful in skin and hair care (1).

Conclusion: HPTLC fingerprint of Rose shows spots of phytochemicals under UV 254 nm and blue and purple bands under white light after derivatization (Fig. **97a**, **b**).

Fig. (97a). Canvas painting of Rose illustrated by Patanjali Research Foundation, Haridwar, India.

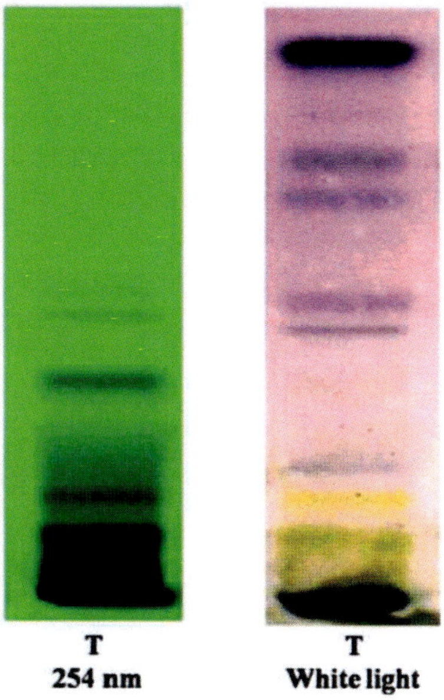

Fig. (97b). HPTLC fingerprint of Rose: T: Rose flower.

Roscoea purpurea Sm (Common Name: Kakoli)

Plant Part: Roots.

Solvent System: Toluene: Ethyl acetate (9: 1 v/v).

Adsorbent: TLC Silica gel $60F_{254}$ (Merck).

Sample Solution Preparation: 500 mg sample dissolved in 10 mL methanol, sonicated and centrifuged. 8 μL of the solution was applied on a TLC plate.

Medicinal Value: Used to treat diarrhoea, fever, inflammation and boost immunity.

Conclusion: HPTLC fingerprint of Kakoli captured at 254 nm, and 366 nm shows numbers of bands at 254 nm and 366 nm. This fingerprint is specific for the sample in given chromatographic conditions (Fig. **98a**, **b**).

Fig. (98a). Canvas painting of Kakoli illustrated by Patanjali Research Foundation, Haridwar, India.

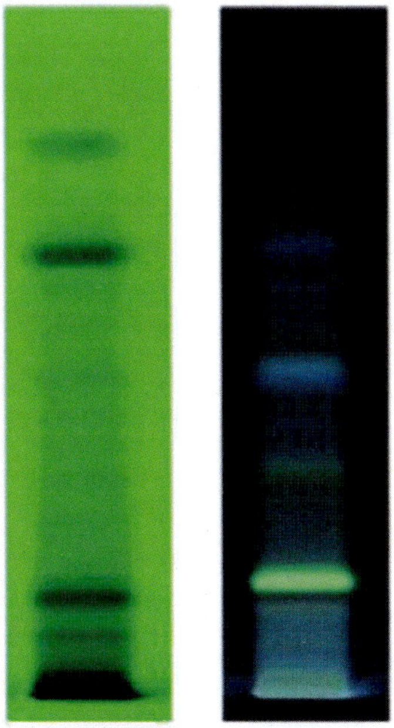

Fig. (98b). HPTLC fingerprint of Kakoli: T: Kakoli roots.

Saraca indica L. (Common Name: Ashok)

Plant Part: Bark.

Solvent System: Toluene: ethyl acetate: formic acid (5: 5: 1 v/v/v).

Adsorbent: TLC Silica gel $60F_{254}$ (Merck).

Sample Solution Preparation: 1000 mg of each sample dissolved in 10 mL of methanol, sonicated and centrifuged. 10 μL of the solution was applied on a TLC plate.

Medicinal Value: Useful in fever, skin allergies and infections and also helpful in irregular menstrual cycle (1).

Conclusion: HPTLC fingerprint of Ashok shows specific bands of phytochemicals at 254 nm and 366 nm and under white light after derivatization (Fig. **99a**, **b**).

Fig. (99a). Canvas painting of Ashok illustrated by Patanjali Research Foundation, Haridwar, India.

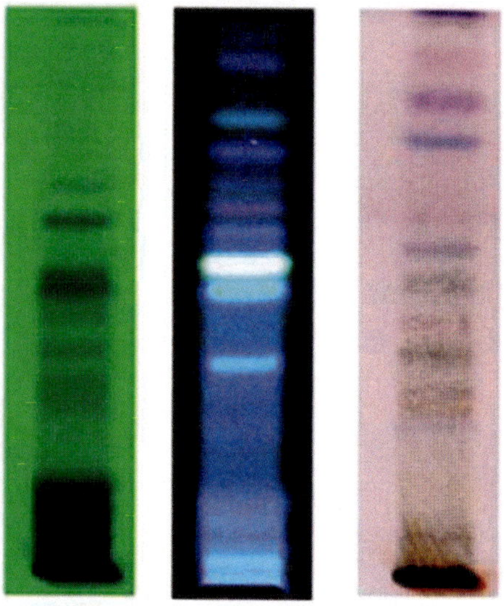

Fig. (99b). HPTLC fingerprint of Ashok: T: Ashok bark.

Sauromatum venosum (Dryand. ex Aiton) Kunth (Common Name: Talia kand)

Plant Part: Root.

Solvent System: Chloroform: methanol: formic acid (8: 2: 1 v/v/v).

Adsorbent: TLC Silica gel $60F_{254}$ (Merck)

Sample Solution Preparation: 1000 mg of the sample dissolved in 10 mL methanol, sonicated and centrifuged. 10µL of the solution was applied on a TLC plate.

Derivatization: Anisaldehyde sulfuric acid.

Medicinal Value: Plant has antibacterial and antioxidant properties.

Conclusion: HPTLC fingerprint of Talia kand represents bands of phytochemicals at 254 nm, and under white light after derivatization (Fig. **100a, b**).

Fig. (100a). Canvas painting of Talia Kand illustrated by Patanjali Research Foundation, Haridwar, India.

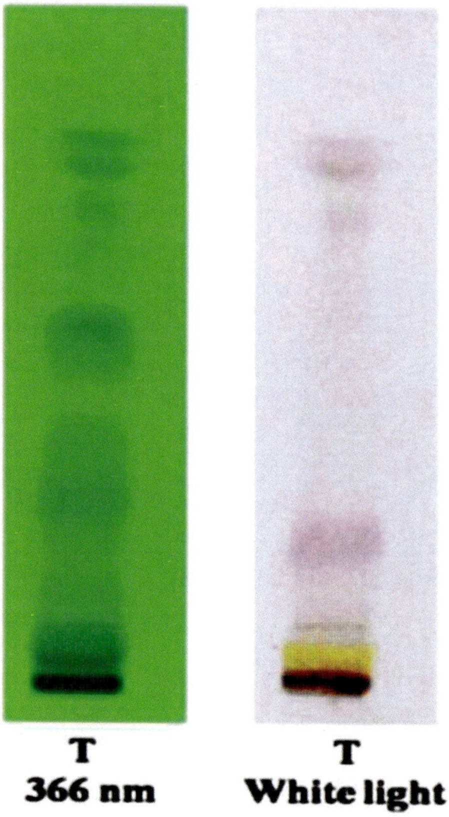

Fig. (100b). HPTLC fingerprint of Talia kand: T: Talia kand root.

Semecarpus anacardium L.f (Common Name: Bhilawa)

Plant Part: Leaves.

Solvent System: Toluene: ethyl acetate: formic acid (10: 10: 2 v/v/v).

Adsorbent: TLC Silica gel $60F_{254}$ (Merck).

Sample Solution Preparation: 500 mg of the sample dissolved in 10 mL methanol, sonicated and centrifuged. 10 µL of the solution was applied on a TLC plate.

Medicinal Value: Used to treat skin diseases, piles and indigestion.

Conclusion: HPTLC fingerprint of Bhilawa represents bands of phytochemicals at 254 nm (Fig. **101a**, **b**).

Fig. (101a). Canvas painting of Bhilawa illustrated by Patanjali Research Foundation, Haridwar, India.

Fig. (101b). HPTLC fingerprint of Bhilawa: T: Bhilawa leaves.

***Senna alexandrina* Mill (Common Name: Senna)**

Plant Part: Leaves.

Solvent System: Toluene: ethyl acetate: formic acid: methanol (10: 10: 1: 5 v/v/v/v).

Adsorbent: TLC Silica gel $60F_{254}$ (Merck).

Standard Solution Preparation: 10 mg of each sennoside A and sennoside B dissolved in 10 mL of methanol. 2 µL of the standard solution was applied on a TLC plate.

Sample Solution Preparation: 500 mg of the sample dissolved in 10 mL methanol sonicated and centrifuged. 8 µL of the solution was applied on the TLC plate.

Medicinal Value: Used to treat constipation, and irritable bowel syndrome (1).

Conclusion: HPTLC image confirms the presence of signature marker sennosoide A and sennoside B at RF 0.21 and 0.16, respectively (Fig. **102a**, **b**).

Fig. (102a). Canvas painting of Senna illustrated by Patanjali Research Foundation, Haridwar, India.

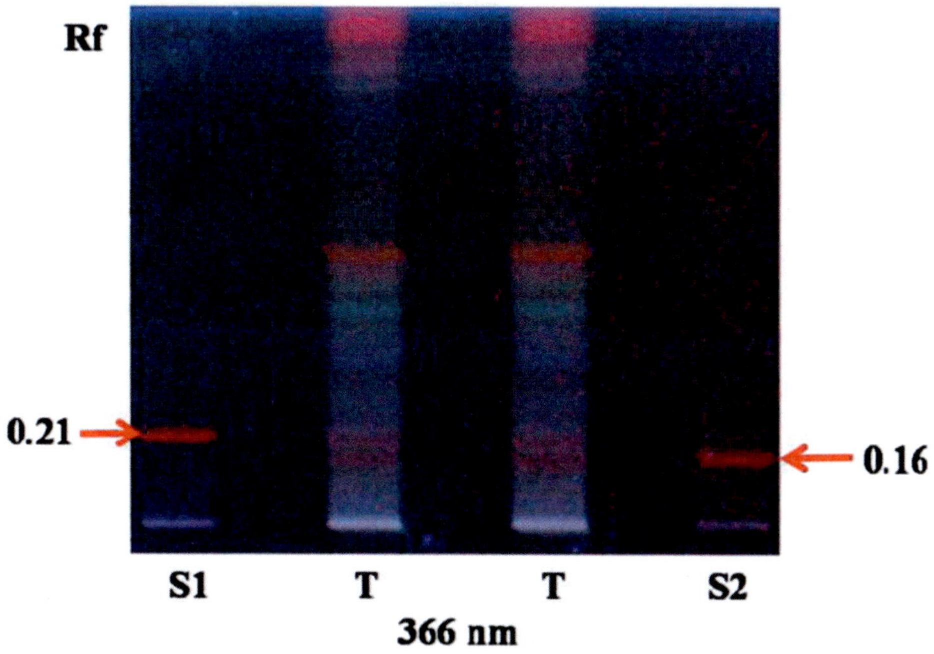

Fig. (102b). HPTLC fingerprint of Senna: S1: Sennoside A, T: Senna leaves, S2: Sennoside B.

***Senna occidentalis* (L.) Link (Common Name: Kasundi)**

Plant Part: Seeds.

Solvent System: Toluene: ethyl acetate: formic acid (10: 10: 2 v/v/v).

Adsorbent: TLC Silica gel $60F_{254}$ (Merck).

Sample Solution Preparation: 500 mg of the sample dissolved in 10 mL methanol, sonicated and centrifuged. 10 μL of the solution was applied on a TLC plate.

Derivatizing Reagent: Anisaldehyde sulfuric acid.

Medicinal Value: The plant has antibacterial, antifungal, anti-malarial, anti-inflammatory, antioxidant, hepatoprotective and immuno suppression activity (1).

Conclusion: HPTLC fingerprint of Kasundi shows bands of phytochemicals at 254 nm and blue and yellow colored bands under white light after derivatization (Fig. **103a**, **b**).

Fig. (103a). Canvas painting of Kasundi illustrated by Patanjali Research Foundation, Haridwar, India.

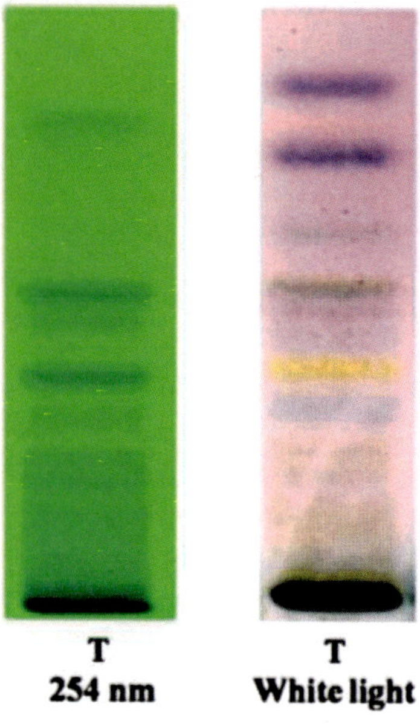

Fig. (103b). HPTLC fingerprint of Kasundi: T: Kasundi seeds.

Senna tora (L.) Roxb (Common Name: Panwad)

Plant Part: Seeds.

Solvent System: Ethyl acetate: toluene: formic acid (7: 3: 1 v/v/v).

Adsorbent: TLC Silica gel $60F_{254}$ (Merck).

Sample Solution Preparation: 1000 mg sample dissolved in 10 mL methanol sonicated and centrifuged. 10 µL of the solution was applied on a TLC plate.

Derivatizing Reagent: Anisaldehyde sulfuric acid.

Medicinal Value: Useful to treat skin ailments, and rheumatic diseases and is also used as laxative.

Conclusion: HPTLC image shows bands of phytochemicals at 254 nm, 366nm and under white light after derivatization (Fig. **104a**, **b**).

Fig. (104a). Canvas painting of Panwad illustrated by Patanjali Research Foundation, Haridwar, India.

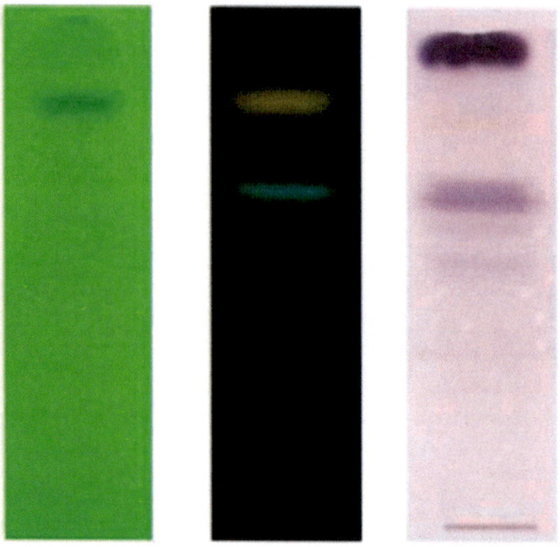

Fig. (104b). HPTLC fingerprint of Panwad: Panwad seeds.

Solanum lycopersicum L. (Common Name: Tamatar)

Plant Part: Fruit.

Solvent System: Hexane: acetone (8: 2 v/v).

Adsorbent: TLC Silica gel $60F_{254}$ (Merck).

Standard Solution Preparation: 10 mg of lycopene was dissolved in 10 mL of methanol. 2 µL of the standard solution was applied on the TLC plate.

Sample Solution Preparation: 500 mg of the sample dissolved in 10 mL methanol sonicated and centrifuged. 8 µL of the solution was applied on a TLC plate.

Medicinal Value: Fruits are rich in lycopene, a substance having beneficial effect on heart and prostate (2).

Conclusion: HPTLC fingerprint of the sample matches with reference standard, showing a specific band of signature molecule lycopene at 0.71 Rf (Fig. **105a**, **b**).

Fig. (105a). Canvas painting of Tomato illustrated by Patanjali Research Foundation, Haridwar, India.

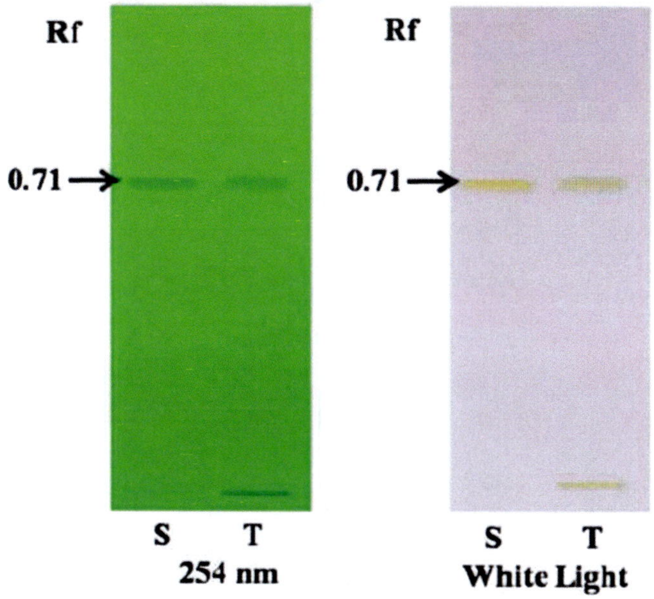

Fig. (105b). HPTLC fingerprinting of Tomato: S: Lycopene, T: Tomato fruit.

Solanum nigrum L. (Common Name: Makoy)

Plant Part: Fruit.

Solvent System: Toluene: ethyl acetate: formic acid (10: 10: 2 v/v/v).

Adsorbent: TLC Silica gel $60F_{254}$ (Merck).

Sample Solution Preparation: 500 mg of the sample dissolved in 10 mL methanol, sonicated and centrifuged. 10 µL of the solution was applied on a TLC plate.

Medicinal Value: Used as tonic, laxative, appetite stimulant, for treating asthma and liver disease (3).

Conclusion: HPTLC fingerprint of Makoy represents bands of phytochemicals at 254 nm and 366 nm (Fig. **106a, b**).

Fig. (106a). Canvas painting of Makoy illustrated by Patanjali Research Foundation, Haridwar, India.

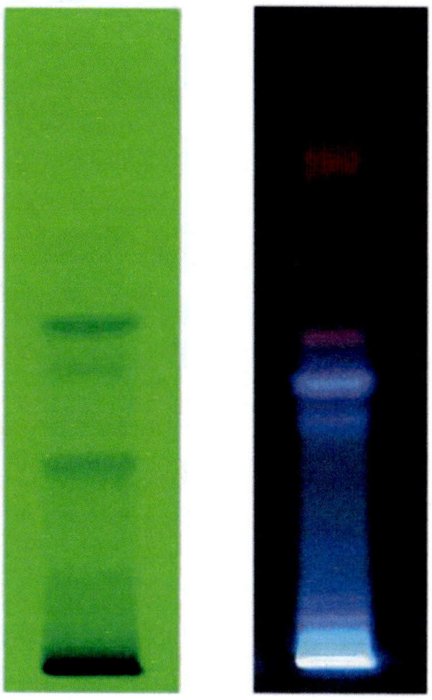

Fig. (106b). HPTLC fingerprint of Makoy: T: Makoy fruit.

Solanum surattense **Burm. F. (Common Name: Chhoti Kateli)**

Plant Part: Aerial part.

Solvent System: Ethyl acetate: toluene: formic acid (5: 5: 0.5 v/v/v).

Adsorbent: TLC Silica gel $60F_{254}$ (Merck).

Standard Solution Preparation: 10 mg of apigenin and caffeic acid was dissolved in 10 mL of methanol. 10 μL of the standard solution was applied on the TLC plate.

Sample Solution Preparation: 500 mg sample dissolved in 10 mL methanol, sonicated and centrifuged.10 μL of the solution was applied on a TLC plate.

Medicinal Value: Used to treat fever, cough, sore throat, and urinary track disorders (1).

Conclusion: HPTLC image confirms the identification of marker apigenin and caffeic acid at Rf 0.45 and 0.52 with respect to standards in Chhoti kateli (Fig. **107a**, **b**).

Fig. (107a). Canvas painting of Chhoti Kateli illustrated by Patanjali Research Foundation, Haridwar, India.

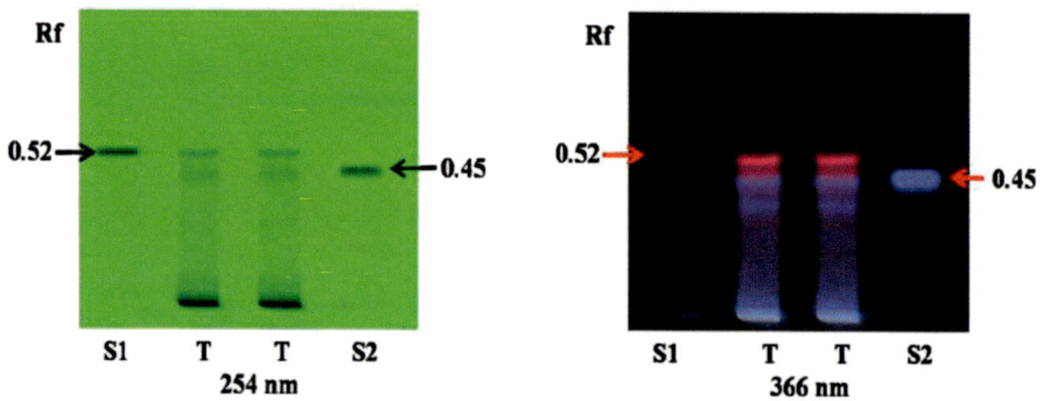

Fig. (107b). HPTLC fingerprint of Chhoti kateli: S1: Apigenin, T: Chhoti kateli, S2: Caffeic acid.

***Swertia chirayita* (Roxb.) H.Karst (Common Name: Chirayita)**

Plant Part: Whole plant.

Solvent System: Chloroform: methanol: formic acid (7: 3: 0.5 v/v/v).

Adsorbent: TLC Silica gel $60F_{254}$ (Merck).

Standard Solution Preparation: 10 mg of lancerin was dissolved in 10 mL of methanol. 10 µL of the standard solution was applied on a TLC plate.

Sample Solution Preparation: 500 mg of sample dissolved in 10 mL methanol, sonicated and centrifuged. 4 µL of the solution was applied on a TLC plate.

Medicinal Value: Used to treat fever and intestinal worms; also have anti-malarial property (2).

Conclusion: HPTLC fingerprint of Chirayita shows the presence of lancerin in the sample at 0.56 Rf (Fig. **108a, b**).

Fig. (108a). Canvas painting of Chirayita illustrated by Patanjali Research Foundation, Haridwar, India.

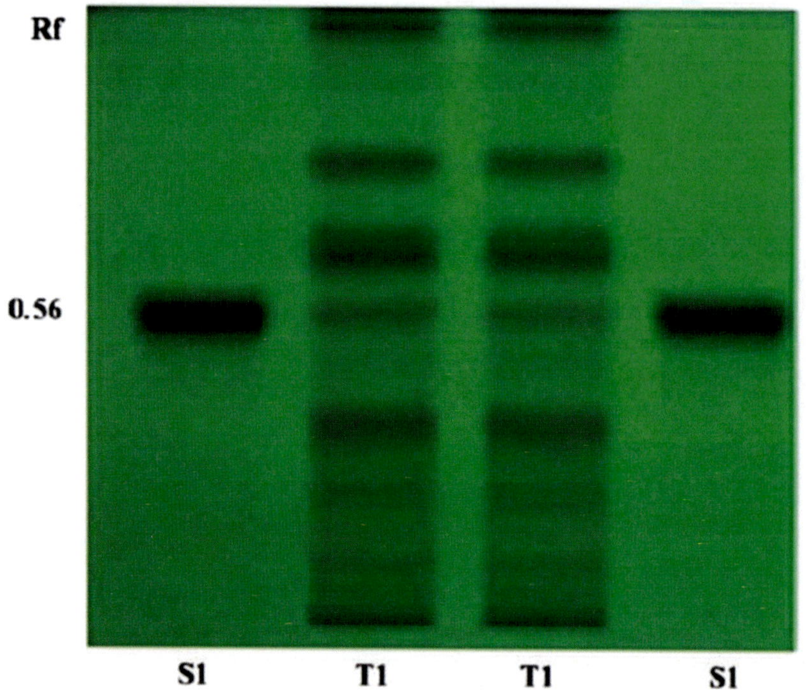

Fig. (108b). HPTLC fingerprint of Chirayita: S1: Lancerin, T1: Chirayita whole plant.

Syzygium aromaticum (L.) Merr. & L.M.Perry (Common Name: Clove)

Plant Part: Buds.

Solvent System: Ethyl acetate: toluene: formic acid (10: 4: 3 v/v/v).

Adsorbent: TLC Silica gel $60F_{254}$ (Merck).

Sample Solution Preparation: 500 mg sample was dissolved in 10 mL methanol, sonicated and centrifuged. 10 μL of the solution was applied on a TLC plate.

Medicinal Value: Useful in the treatment of vomiting flatulence, nausea, liver, bowel diseases and stomach disorders (3).

Conclusion: HPTLC fingerprint of Clove shows a specific banding pattern at 254 nm and 366 nm (Fig. **109a, b**).

Fig. (109a). Canvas painting of Clove illustrated by Patanjali Research Foundation, Haridwar, India.

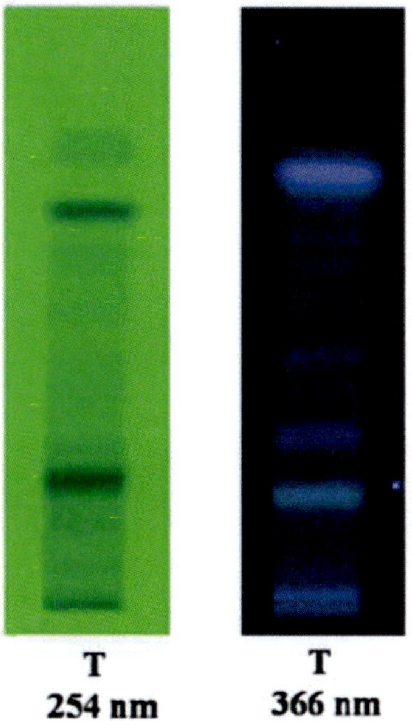

Fig. (109b). HPTLC fingerprint of Clove: T: Clove buds.

Terminalia arjuna (Roxb. ex DC.) Wight & Arn (Common Name: Arjun)

Plant Part: Bark.

Solvent System: Ethyl acetate: toluene: formic acid: acetic acid (5: 3: 0.5: 1 v/v/v).

Adsorbent: TLC Silica gel $60F_{254}$ (Merck).

Standard Solution Preparation: 10 mg of arjunic acid was dissolved in 10 mL of methanol. 10 µL of the standard solution was applied on a TLC plate.

Sample Solution Preparation: 500 mg sample dissolved in 10 mL methanol sonicated and centrifuged.10 µL of the solution was applied on a TLC plate.

Derivatizing Reagent: Anisaldehyde sulfuric acid.

Medicinal Value: The plant has cardiovascular, antioxidant and anti-inflammatory properties (1).

Conclusion: HPTLC fingerprint of Arjun shows the presence of arjunic acid at Rf 0.70 with respect to the standard (Fig. **110a**, **b**).

Fig. (110a). Canvas painting of Arjun illustrated by Patanjali Research Foundation, Haridwar, India.

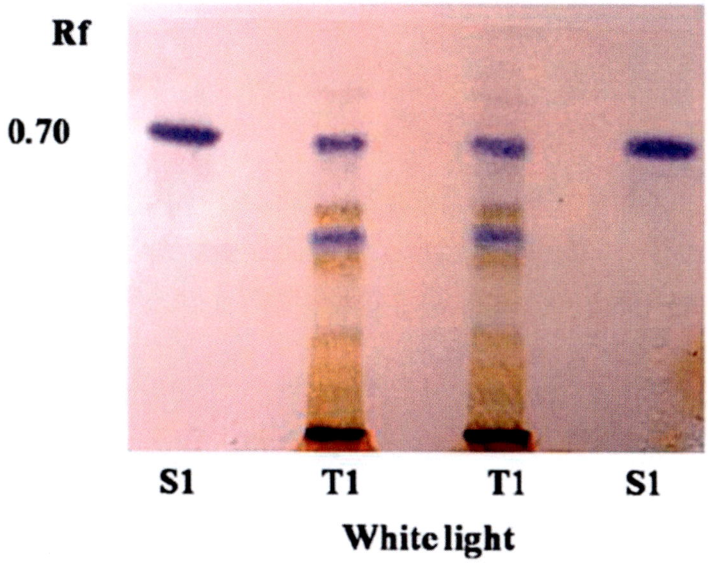

Fig. (110b). HPTLC fingerprint of Arjun: S1: Arjunic acid, T1: Arjun bark.

Terminalia bellirica **(Gaertn.) Roxb (Common Name: Bahera)**

Plant Part: Fruit.

Solvent System: Ethyl acetate: toluene: formic acid (5: 2: 1.5 v/v/v).

Adsorbent: TLC Silica gel $60F_{254}$ (Merck).

Standard Solution Preparation: 10 mg of gallic acid was dissolved in 10 mL of methanol. 20 µL of the standard solution was applied on a TLC plate.

Sample Solution Preparation: 1000 mg sample dissolved in 10 mL methanol sonicated and centrifuged. 6 µL of the solution was applied on a TLC plate.

Medicinal Value: Used in skin disorder, fever, indigestion and cough.

Conclusion: HPTLC image confirms the identification of signature marker gallic acid at Rf 0.61, with respect to standard (Fig. **111a**, **b**).

Fig. (111a). Canvas painting of Bahera illustrated by Patanjali Research Foundation, Haridwar, India.

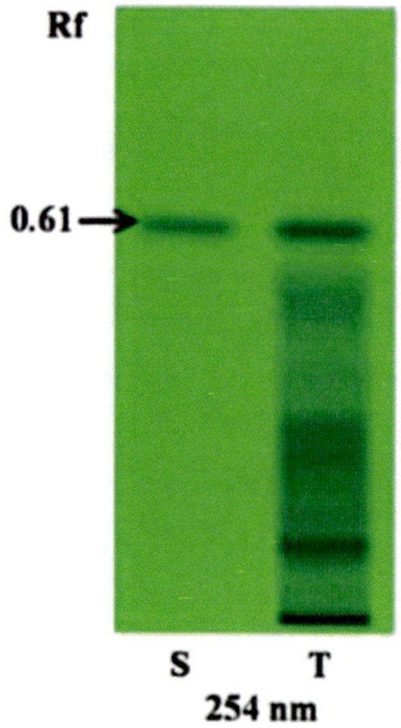

Fig. (111b). HPTLC fingerprint of Bahera: S: Gallic acid, T: Bahera fruit.

Terminalia chebula Retz (Common Name: Harad)

Plant Part: Harad.

Solvent System: Ethyl acetate: toluene: formic acid (5: 2: 1.5 v/v/v).

Adsorbent: TLC Silica gel $60F_{254}$ (Merck).

Standard Solution Preparation: 10 mg of gallic acid and chebulinic acid was dissolved in 10 mL of methanol. 20 µL of the standard solution was applied on a TLC plate.

Sample Solution Preparation: 1000 mg sample dissolved in 10 mL methanol, sonicated and centrifuged. 6 µL of the solution was applied on a TLC plate.

Medicinal Value: Used in skin disorder, fever, indigestion and cough (3).

Conclusion: HPTLC image confirms the identification of signature marker gallic acid and chebulinic acid at Rf 0.61 and 0.20 respectively, with respect to standards (Fig. **112a**, **b**).

Fig. (112a). Canvas painting of Harad illustrated by Patanjali Research Foundation, Haridwar, India.

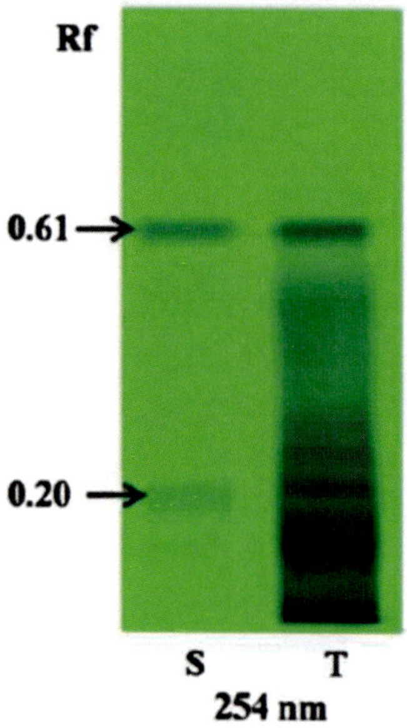

Fig. (112b). HPTLC fingerprint of Harad: S: Gallic acid (0.61) and chebulinic acid (0.20), T: Harad fruit.

Tinospora cordifolia (Willd.) Hook.f. & Thomson (Common Name: Giloy)

Plant Part: Stem.

Solvent System: Ethyl acetate: formic acid: acetic acid: water (10: 1.1: 1.1: 2.6 v/v/v/v).

Adsorbent: TLC Silica gel $60F_{254}$ (Merck).

Standard Solution Preparation: 10 mg of magnoflorine was dissolved in 10 mL of methanol. 10 µL of the standard solution was applied on a TLC plate.

Sample Solution Preparation: 500 mg sample dissolved in 10 mL methanol: water (70: 30 v/v) sonicated and centrifuged. 4 µL of the solution was applied on a TLC plate.

Medicinal Value: Boosts immunity, treats chronic fever, improves digestion, purifies blood, and reduces stress and anxiety (1).

Conclusion: Rf 0.28 of the sample confirms the presence of magnoflorine in sample by HPTLC fingerprint (Fig. **113a**, **b**).

Fig. (113a). Canvas painting of Giloy illustrated by Patanjali Research Foundation, Haridwar, India.

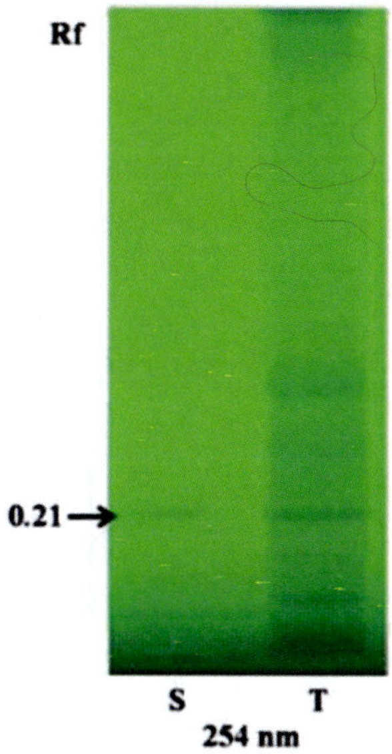

Fig. (113b). HPTLC fingerprint of Giloy: S: Magnoflorine, T: Giloy stem.

Trachyspermum ammi (L.) Sprague (Common Name: Carom)

Plant Part: Seeds.

Solvent System: Toluene: ethyl acetate: formic acid: acetic acid (7: 3: 1 v/v/v).

Adsorbent: TLC Silica gel $60F_{254}$ (Merck).

Sample Solution Preparation: 1000 mg sample dissolved in 10 mL methanol, sonicated and centrifuged. 10 µL of the solution was applied on a TLC plate.

Derivatizing Reagent: Anisaldehyde sulfuric acid.

Medicinal Value: Improves the flow of stomach acid, which can help relive indigestion, bloating and gas (1).

Conclusion: HPTLC fingerprint of Carom shows specific bands of phytochemicals at 254 nm, 366 nm and under white light after derivatization (Fig. **114a, b**).

Fig. (114a). Canvas painting of Carom illustrated by Patanjali Research Foundation, Haridwar, India.

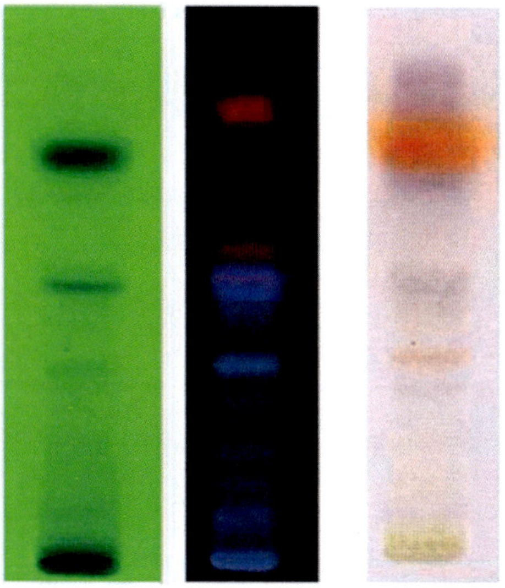

Fig. (114b). HPTLC fingerprint of Carom: T: Carom seeds.

Trigonella foenum-graecum L. (Common Name: Fenugreek)

Plant Part: Seeds.

Solvent System: Ethyl acetate: methanol: water: formic acid (10: 8: 4.6: 1.6 v/v/v/v).

Adsorbent: TLC Silica gel $60F_{254}$ (Merck).

Standard Solution Preparation: 10 mg of trigonelline was dissolved in 10 mL of methanol. 2 μL of the standard solution was applied on a TLC plate.

Sample Solution Preparation: 500 mg of the sample dissolved in 10 mL methanol, sonicated and centrifuged. 10 μL of the solution was applied on a TLC plate.

Medicinal Value: Useful in body ache, diarrhoea, diabetes, hair care and skin care (3).

Conclusion: Rf 0.37 in HPTLC fingerprint showing the presence of trigonelline in Fenugreek (Fig. **115a**, **b**).

Fig. (115a). Canvas painting of Fenugreek illustrated by Patanjali Research Foundation, Haridwar, India.

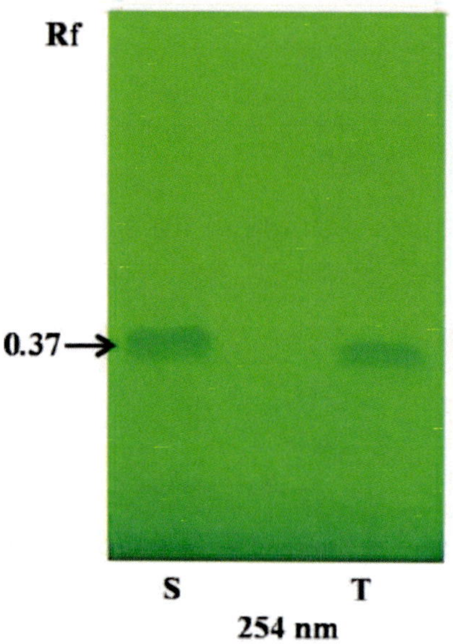

Fig. (115b). HPTLC fingerprint of Fenugreek: S: Trigonelline, T: Fenugreek seeds.

Urtica ardens Link (Common Name: Stinging Nettle)

Plant Part: Leaves.

Solvent System: Ethyl acetate: toluene: formic acid (7: 3: 1 v/v/v).

Adsorbent: TLC Silica gel $60F_{254}$ (Merck).

Sample Solution Preparation: 500 mg sample dissolved in 10 mL water, sonicated and centrifuged. 10 µL of the solution was applied on a TLC plate.

Derivatizing Reagent: Anisaldehyde sulfuric acid.

Medicinal Value: Useful to treat urinary track disorders and reduce inflammation. It is rich in antioxidants, antimicrobial and diuretic.

Conclusion: HPTLC fingerprint of Stinging nettle shows the band of phytochemical under 254 nm 366 nm and under white light after derivatization (Fig. **116a**, **b**).

Fig. (116a). Drawing of Stinging Nettle illustrated by Patanjali Research Foundation, Haridwar, India.

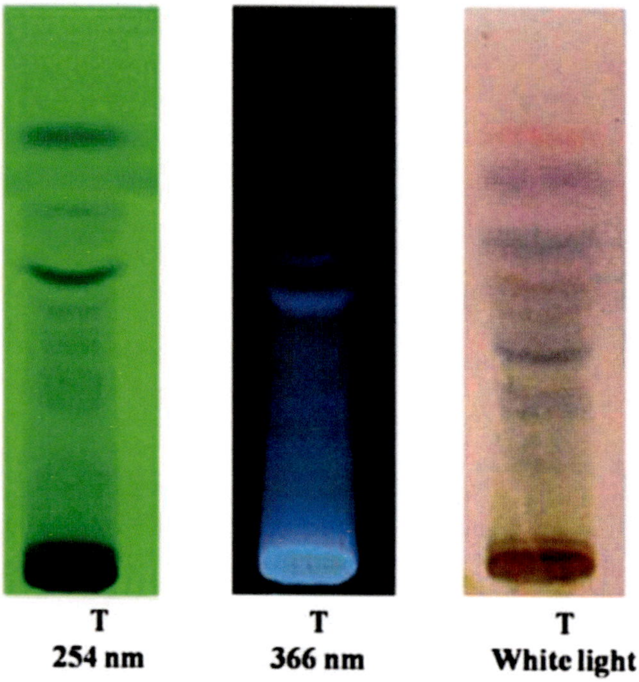

Fig. (116b). HPTLC fingerprint of Stinging nettle: T: Stinging nettle leaves.

Viola odorata L. (Common Name: Sweet Violet)

Plant Part: Leaves.

Solvent System: Hexane: ethyl acetate (8: 2 v/v).

Adsorbent: TLC Silica gel $60F_{254}$ (Merck).

Sample Solution Preparation: 1000 mg sample dissolved in 10 mL methanol sonicated and centrifuged. 10 µL of the solution was applied on a TLC plate.

Derivatizing Reagent: Anisaldehyde sulfuric acid.

Medicinal Value: Used for the treatment of headache, cough colds, bronchitis, and fever (3).

Conclusion: HPTLC fingerprint of Sweet violet shows a specific banding pattern at 254 nm, 366 nm and under white light after derivatization (Fig. **117a**, **b**).

Fig. (117a). Canvas painting of Sweet violet illustrated by Patanjali Research Foundation, Haridwar, India.

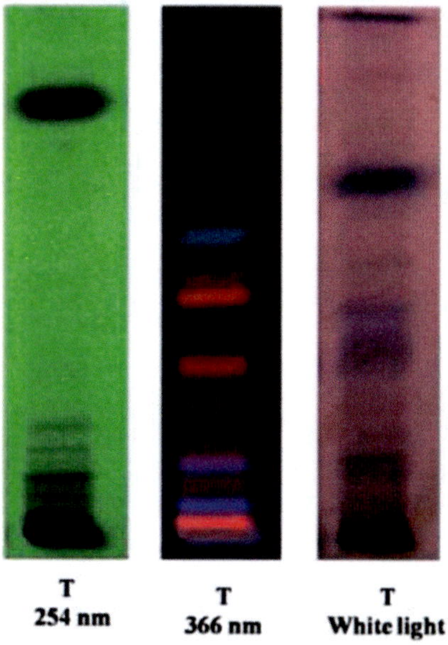

Fig. (117b). HPTLC fingerprint of Sweet violet: T: Sweet violet leaves.

Vitex negundo L. (Common Name: Nirgundi)

Plant Part: Whole plant.

Solvent System: Ethyl acetate: acetic acid: water (8: 1: 1 v/v/v).

Adsorbent: TLC Silica gel $60F_{254}$ (Merck).

Sample Solution Preparation: 500 mg of the sample dissolved in 10 mL methanol, sonicated and centrifuged. 8 μL of the solution was applied on a TLC plate.

Medicinal Value: It has analgesic, antipyretic, diuretic and anti-inflammatory properties (2).

Conclusion: HPTLC fingerprint of Nirgundi shows a specific banding pattern at 254 nm and 366 nm (Fig. **118a, b**).

Fig. (118a). Canvas painting of Nirgundi illustrated by Patanjali Research Foundation, Haridwar, India.

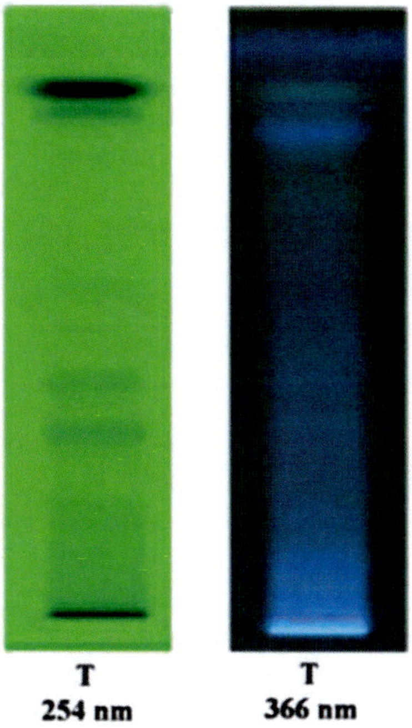

Fig. (118b). HPTLC fingerprint of Nirgundi: T: Nirgundi whole plant.

Vitis vinifera L. (Common Name: Grapes)

Plant Part: Seeds.

Solvent System: Ethyl acetate: toluene: formic acid (5: 5: 0.5 v/v/v).

Adsorbent: TLC Silica gel $60F_{254}$ (Merck).

Sample Solution Preparation: 500 mg sample dissolved in 10 mL methanol sonicated and centrifuged. 10 μL of the solution was applied on a TLC plate.

Derivatization: Anisaldehyde sulfuric acid.

Medicinal Value: Beneficial for eyes, skin and hair. Promotes longevity, and lessens inflammation.

Conclusion: HPTLC fingerprint represents bands of phytochemicals at 254 nm, 366nm and under white light (Fig. **119a**, **b**).

Fig. (119a). Canvas painting of Grapes illustrated by Patanjali Research Foundation, Haridwar, India.

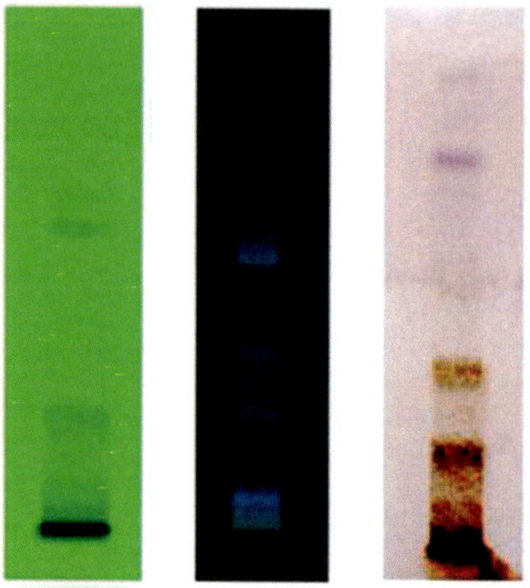

Fig. (119b). HPTLC fingerprint of Grapes: T: Grapes seeds.

Withania somnifera (L.) Dunal (Common Name: Ashwagandha)

Plant Part: Roots.

Solvent System: Ethyl acetate: toluene: formic acid (5: 5: 1 v/v/v).

Adsorbent: TLC Silica gel $60F_{254}$ (Merck).

Standard Solution Preparation: 10 mg of each standard withaferine A, withanolide A and withanolide B, was dissolved in 10 mL of methanol. 5 µL of each standard solution was applied on a TLC plate.

Sample Solution Preparation: 500 mg sample dissolved in 10 mL methanol: water (60: 40 v/v), sonicated and centrifuged. 8 µL of the solution was applied on a TLC plate.

Medicinal Value: Immunity booster, rejuvenator, aphrodisiac, reduce stress and anxiety (1).

Conclusion: Photograph of the TLC plate validating the presence of withaferine A, withanolide A and withanolide B in Aahwagandha. The sample showing specific bands of markers at Rf values 0.28, 0.38 and 0.51 with respect to the standards (Fig. **120a**, **b**).

Fig. (120a). Canvas painting of Ashwagandha illustrated by Patanjali Research Foundation, Haridwar, India.

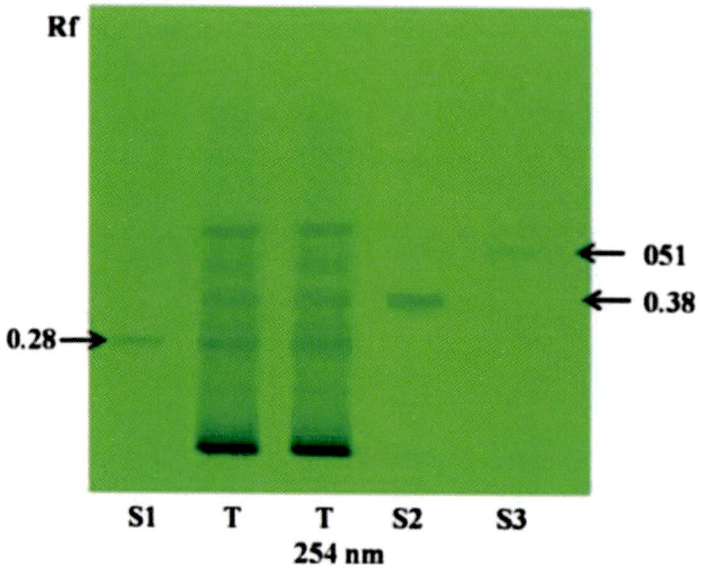

Fig. (120b). HPTLC fingerprint of Ashwagandha: S1: Withaferine A, T: Ashwagandha roots, S2: Withanolide A, S3: Withanolide B.

Zanthoxylum armatum (Thunb.) Druce (Common Name: Tambru)

Plant Part: Seeds.

Solvent System: Toluene: ethyl acetate formic acid (8: 2: 0.3 v/v/v).

Adsorbent: TLC Silica gel $60F_{254}$ (Merck).

Sample Solution Preparation: 500 mg of the sample dissolved in 10 mL methanol, sonicated and centrifuged. 10 µL of the solution was applied on a TLC plate.

Medicinal Value: It is anti-bacterial and antifungal, and helps to relieve toothache.

Conclusion: HPTLC fingerprint of Tambru showing a specific pattern of bands at 254 nm and fluorescent bands at 366 nm (Fig. **121a**, **b**).

Fig. (121a). Canvas painting of Tambru illustrated by Patanjali Research Foundation, Haridwar, India.

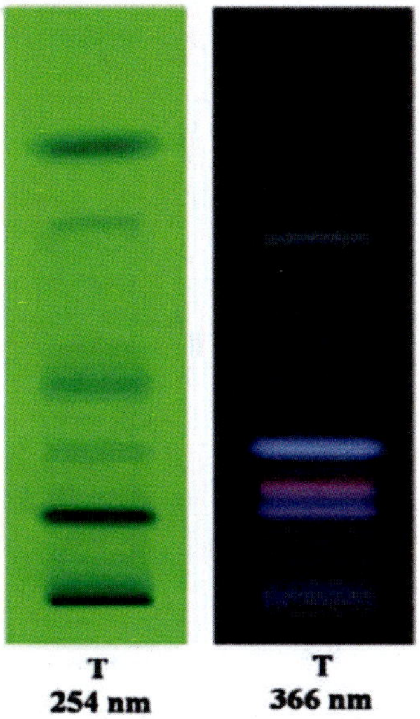

Fig. (121b). HPTLC fingerprint of Tambru: T: Tambru seeds.

Zingiber officinale Roscoe (Common Name: Ginger)

Plant Part: Rhizome.

Solvent System: Toluene: ethyl acetate: diethyl ether (6: 3: 1 v/v/v).

Adsorbent: TLC Silica gel $60F_{254}$ (Merck).

Standard Solution Preparation: 10 mg of each, 6 gingerol and 8 gingerol was dissolved in 10 mL of methanol. 5 µL of each standard solution was applied on a TLC plate.

Sample Solution Preparation: 500 mg sample dissolved in 10 mL methanol, sonicated and centrifuged. 8 µL solution was applied on a TLC plate.

Derivatization: Anisaldehyde sulfuricacid.

Medicinal Value: Useful in cough, and cold. It improves digestion, effective in relieving dizziness, menstrual pain and arthritis pain (3).

Conclusion: HPTLC fingerprint validates the presence of 6 gingerol and 8 gingerol in Ginger. The sample showing bands of markers at Rf values 0.55 and 0.60 with respect to the standards (Fig. **122a**, **b**).

Fig. (122a). Canvas painting of Ginger illustrated by Patanjali Research Foundation, Haridwar, India.

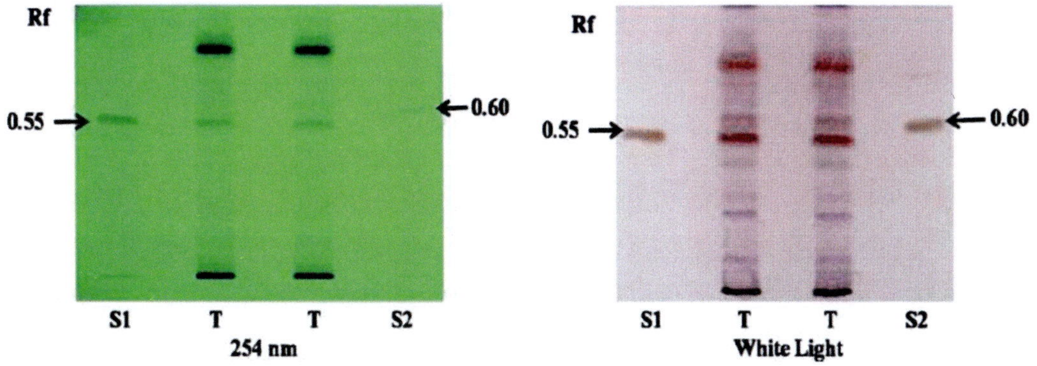

Fig. (122b). HPTLC fingerprint of Ginger: S1: 6 Gingerol, T: Ginger rhizome, S2: 8 Gingerol.

CONCLUDING REMARKS

HPTLC technique has been long in use for the purpose of phytochemical identification. However, a systematic compilation of process parameters for identifying different plants of medicinal Value was missing. This chapter precisely overcomes that technical shortcoming with credentials to serve as an analytical guide for the identification and characterization of medicinal plants.

APPENDIX

Solvent Systems for Screening of Phytochemical Compounds of Herbal Plants by TLC/ HPTLC

Table A1. Solvent system for screening of phytochemical compounds of herbal plants by TLC/HPTLC.

Class of Compound	Mobile Phase	Derivatization
Alkaloids	Toluene: ethyl acetate: diethyl amine (7: 2: 1 v/v/v)	Dragendroff's reagent
	Toluene: methanol: diethyl amine (8: 1: 1 v/v/v)	
	Toluene: ethyl acetate: methanol: ammonia (25%) (3: 3: 1.5: 0.1 v/v/v/v)	
	Butanol: acetic acid: water (6: 2: 1 v/v/v)	
Essential oils	Toluene: ethyl acetate (9.3: 0.7 v/v)	Vanillin sulfuric acid
Flavonoids	Ethyl acetate: formic acid: glacial acetic acid: water (10: 1.1: 1.1: 2.3 v/v/v/v)	10 % methanolic sulfuric acid / Natural product reagent
	Toluene: ethyl acetate: formic acid (5: 5: 1 v/v/v)	
	Tetra hydrofuran: toluene: formic acid: water (1.6: 0.8: 0.2: 0.1 v/v/v/v)	
Glycosides	Ethyl acetate: methanol: water (20: 2.8: 2 v/v/v)	Alcoholic potassium hydroxide
Saponins	Chloroform: acetic acid: methanol: water (6.4: 3.2: 1.2: 0.8 v/v/v/v)	Anisaldehyde sulfuric acid
	Butanol: water: acetic acid (5: 4: 1 v/v/v)	
Steroids	Butanol: methanol: water (3: 1: 1 v/v/v)	Anisaldehyde sulfuric acid
Sterols	Ethyl acetate: chloroform (6: 4 v/v)	Anisaldehyde sulfuric acid
	Chloroform: methanol (9: 1 v/v)	

Acharya Balkrishna

Appendix I

Derivatization Reagents for Detection of Secondary Metabolites of Herbal Plants by TLC/HPTLC

Table A2. Derivatization reagents for detection of secondary metabolites of herbal plants by TLC/HPTLC.

Name of Reagent	Procedure	Examination	Detection of
Alcoholic iron chloride	2 g iron chloride is dissolved in 10 mL water and diluted to 200 mL with ethanol.	White light	Alkaloids, flavonoids, plant acids, phenols and tannins.
Alcoholic potassium hydroxide	5-10 g potassium hydroxide is dissolved in 10 mL of water and diluted to 100 mL with ethanol.	White light, UV 366 nm	Anthraquinones, coumarins and flavonoids.
Anisaldehyde sulfuric acid	Place 170 mL of methanol in 200 mL glass bottle and cool it down in water-ice cube bath. To the ice cold methanol, slowly and carefully add 20 mL glacial acetic acid and 10 mL sulphuric acid. Allow it to cool to room temperature, then add 1 mL para-anisaldehyde.	White light	Sterols, terpenoides and Saponins
Copper sulphate	20 g copper sulphate is dissolved in 180 mL water containing 16 mL ortho-phosphoric acid (80% in water).	White light	Lipids
Dregendroff's reagent	Solution A- 0.85 g basic bismuth nitrate is dissolved in 10 mL acetic acid and 40 mL water under heating. Solution B- 8 g potassium iodide is dissolved in 30 mL water. Just before spraying, 1 mL of each solution is mixed with 4 mL acetic acid and 20 mL water.	White light	Alkaloids
Natural product reagent	Solution A- 1 g of diphenylborinic acid aminoethylester is dissolved in 100 mL methanol. Solution B 5 g of polyethylene glycol 400 (macrogol) is dissolved in 100 mL ethanol.	UV 366	Flavonoids, carbohydrates, anthocyanines, plant acids
Ninhydrin reagent	0.6 g ninhydrin is dissolved in 190 mL isopropanol, then 10 mL acetic acid is added.	White light	Amino acid, ephedrins

(Table A2) cont.....

Name of Reagent	Procedure	Examination	Detection of
Vanillin sulphuric acid	1 g of vanillin is dissolved in 100 mL of 96 % ethanol add 2 mL of concentrated sulphuric acid.	White light, UV 366 nm	Terpenoids, ergot alkaloids, sterols

Appendix II

Table A3. List of medicinal plants.

S. No.	Botanical Name	Hindi Name	Sanskrit Name (Phonetics)
1	*Achyranthes aspera* L.	Apamarga	Apāmārgakaḥ kharamañjikaḥ
2	*Acorus calamus* L.	Vach	Ugragandhakaḥ vacaḥ
3	*Aegle marmelos (L.) Corrêa*	Bael	Bilvakaḥ tripatraḥ
4	*Albizia odoratissima* (L.f.) Benth.	Siris	Mṛdusumakam jāliśimbam
5	*Allium sativum* L.	Lehsun	Gandhālukaḥ rasonaḥ
6	*Aloe vera (L.) Burm.f.*	Aloe Vera	Majjapatrakam aṅgārasumam
7	*Anacyclus pyrethrum* (L.) Lag.	Akarkara	Ākārakarabhakaḥ chinnapatraḥ
8	*Andrographis paniculata* (Burm.f.) Nees	Kalmegh	Kālameghakaḥ tiktayavaḥ
9	*Areca catechu* L.	Supari	Pūgaphalakam vartulam
10	*Asparagus racemosus* Willd.	Shatavar	Śatāvarakaḥ adharakaṇṭaḥ
11	*Azadirachta indica* A.Juss.	Neem	Picumardakaḥ nimbaḥ
12	*Baccharoides anthelmintica* (L.) Moench	Kali jeera	Gandhimūlakam Kṛmighnam
13	*Bacopa monnieri* (L.) Wettst.	Brahmi	Brāhmikā medhyā
14	*Bauhinia variegata* L.	Kachnar	Yugmapatrakam bhinnadalam
15	*Berberis aristata* DC.	Daruhaldi	Pītadrukaḥ haridruḥ
16	*Berberis napaulensis* (DC.) Spreng.	Mahonia	Pītadrukaḥ daravalkaḥ
17	*Boerhavia diffusa* L.	Punarnava	Punarnavakaḥ raktakāṇḍaḥ
18	*Brassica juncea* (L.) Czern.	Lal sarson	Sarṣapakaḥrājikaḥ
19	*Cannabis sativa* L.	Bhang	Bhaṅgikā pītasapatrā
20	*Capsicum frutescens* L.	Mirch	Kaṭuvīrakā aṇḍapatrā
21	*Carica papaya* L.	Papita	Kumbhakarkaṭikā kṛṣṇerūḥ
22	*Carissa carandas* L.	Karonda	Karamardakaḥ pravṛttaphalaḥ
23	*Carum carvi* L.	Shah jeera	Jīrakam kapilam
24	*Cassia fistula* L.	Amaltas	Hemapuṣpakam kṛtamālam

(Table A3) cont.....

S. No.	Botanical Name	Hindi Name	Sanskrit Name (Phonetics)
25	*Catharanthus roseus* (L.) G.Don	Sadabhar	Sadāpuṣpakam pāṭalākṣam
26	*Cedrus deodara* (Roxb. ex D.Don) G.Don	Deodar	Devadrujaḥ dṛḍhakāṣṭhaḥ
27	*Celastrus paniculatus* Willd.	Malkangani	Kākāṇḍakaḥ gucchikramaḥ
28	*Chlorophytum borivilianum* Santapau & R.R.Fern.	Safed musli	Sitamuśalikā bhallapatrā
29	*Cinnamomum camphora* (L.) J.Presl	Kapoor	Gandhajātakam karpūram
30	*Cinnamomum tamala* (Buch.-Ham.) T.Nees & C.H.Eberm.	Tejpatta	Gandhajātakam tamālam
31	*Cinnamomum verum* J.Presl	Dalchini	Gandhajātakam tvak
32	*Cissus quadrangularis* L.	Hadjod	Vajravalī asthisaṁhārī
33	*Cocculus hirsutus* (L.) W.Theob.	Jaljamini	Śapherukaḥ mahāmūlaḥ
34	*Commiphora mukul* (Hook. ex Stocks) Engl.	Guggul	Guggulakaḥ devadhūpaḥ
35	*Convolvulus prostratus* Forssk.	Shankhpushpi	Kambumālakaḥ śaṅkhapuṣpaḥ
36	*Cordia dichotoma* G.Forst.	Lisoda	Śleṣmātakaḥ śeluḥ
37	*Coriandrum sativum* L.	Dhania	Dhāneyakaḥ kustumburuḥ
38	*Cornus capitata* Wall.	Thanboi	Cirāṅgaphalakam carmipatram
39	*Cressa cretica* L.	Rudanti	Rudantakaḥ caṇapatraḥ
40	*Crinum latifolium L.*	Sudarshan	Sudarśanakaḥ urupatraḥ
41	*Cullen corylifolium* (L.) Medik.	Bakuchi	Indulekhikā bākucī
42	*Cuminum cyminum* L.	Jeera	Ajājikā jīrakā
43	*Curcuma longa* L.	Haldi	Haridrakā pītāṅgā
44	*Cyperus scariosus* R.Br.	Nagarmotha	Jalamuk nāgaramustakaḥ
45	*Datura stramonium* L.	Datura	Dhattūrakaḥ kṛṣṇḥ
46	*Daucus carota* L.	Gajar	Garjarakaḥ nāraṅgaḥ
47	*Diplocyclos palmatus* (L.) C.Jeffrey	Shivlingi	Śivaliṅgikā karapatrā
48	*Eclipta prostrata* (L.) L.	Bhringraj	Bhṛṅgarājakaḥ keśyaḥ
49	*Elaeocarpus serratus L.*	Rudraksha	Rudrākṣḥ ārīpatraḥ
50	*Elettaria cardamomum* (L.) Maton	Elaichi	Elikā truṭiḥ
51	*Embelia ribes* Burm.f.	Viavidang	Viḍaṅgakaḥ kṛmighnaḥ
52	*Foeniculum vulgare* Mill.	Saunf	Miśreyakā śatāhvā
53	*Glycyrrhiza glabra* L.	Mulethi	Yaṣṭimadhukaḥ aroma
54	*Habenaria intermedia* D.Don	Riddhi	Ṛddhikārcī tuṣṭiḥ
55	*Hemidesmus indicus* (L.) R.Br.	Anantmool	Sārivakā sitaśirā

(Table A3) cont.....

S. No.	Botanical Name	Hindi Name	Sanskrit Name (Phonetics)
56	*Holarrhena pubescens* Wall. ex G.Don	Indrajao (Seeds)	Kuṭajakaḥ yavaphalaḥ
57	*Holarrhena pubescens* Wall. ex G.Don	Kutaj (Bark)	Kuṭajakaḥ yavaphalaḥ
58	*Lagenaria siceraria* (Molina) Standl.	Lauki	Kūpyālābukā pratumbī
59	*Lavandula stoechas* L.	Ustekhaddus	Phullakā rājisumā
60	*Lawsonia inermis* L.	Henna	Madayantikā rañjanā
61	*Leucas cephalotes* (Roth) Spreng.	Dronpushpi	Droṇapuṣpakam śīrṣakramam
62	*Lilium polyphyllum* D.Don	Ksheer kakoli	Lalāmakaḥ bahupatraḥ
63	*Mentha × piperita L.*	Mint	Pūtihākā tailapatrā
64	*Momordica charantia* L.	Karela	Kāravellakaḥ dīrghaphalaḥ
65	*Moringa oleifera* Lam.	Sahjan	Śigrukaḥ sitapuṣpaḥ
66	*Murraya paniculata* (L.) Jack	Gandhani	Kaiḍaryakaḥ pūgapuṣpaḥ
67	*Myristica fragrans* Houtt.	Javatri (Aril)	Gandhacolakaḥ raktaḥ
68	*Myristica fragrans* Houtt.	Jaiphal (Fruit)	Gandhacolakaḥ raktaḥ
69	*Nardostachys jatamansi* (D.Don) DC.	Jatamansi	Jaṭālakaḥ naladaḥ
70	*Nymphaea nouchali* Burm.f.	Neel kamal	Kumudakam nīlam
71	*Ocimum basilicum* L.	Tulsi	Sumañjarikā śyāmā
72	*Ocimum tenuiflorum* L.	Tulsi	Sumañjarikā rāmā
73	*Operculina turpethum* (L.) Silva Manso	Nisoth	Pīnaśalkam tāmrakali
74	*Phyllanthus emblica* L.	Amla	Āmalakaḥ dhātrīphalaḥ
75	*Picrorhiza kurroa* Royle ex Benth.	Kutki	Kaṭukā tiktā
76	*Piper longum* L.	Chhoti pippali	Kaṇikā pippalī
77	*Piper nigrum* L.	Marich	Kaṇikā maricā
78	*Piper retrofractum* Vahl	Badi pippali	Kaṇikā golasapatrā
79	*Pistacia chinensis subsp. integerrima (J.L.Stewart) Rech.f.*	Kakra singhi	Karkaṭaśṛṅgakam raktapallavam
80	*Putranjiva roxburghii* Wall.	Putrajeevak	Putraṇjīvakaḥ tigmāntaḥ
81	*Raphanus raphanistrum subsp. sativus (L.) Domin*	Mooli	Mūlakam hastidantam
82	*Rheum emodi Wall.*	Revandchini	Gucchikramakaḥ pītamūlaḥ
83	*Rosa indica* L.	Gulab	Taruṇikā pāṭalā
84	*Roscoea purpurea* Sm.	Kakoli	Kākolikā vṛntākā
85	*Saraca indica* L.	Ashok	Tāmrapallavakaḥ aṅgārakaḥ

(Table A3) cont.....

S. No.	Botanical Name	Hindi Name	Sanskrit Name (Phonetics)
86	*Sauromatum venosum* (Dryand. ex Aiton) Kunth	Talia kand	Saraṭāṅkaḥ vṛntākaḥ
87	*Semecarpus anacardium* L.f.	Bhilawa	Bhallātakaḥ sphoṭalaḥ
88	*Senna alexandrina* Mill.	Senna	Cakramardakaḥ tanupatraḥ
89	*Senna occidentalis* (L.) Link	Kasundi	Cakramardakaḥ nāraṅgaśiraḥ
90	*Senna tora* (L.) Roxb.	Panwad	Cakramardakaḥ aṇḍadalaḥ
91	*Solanum lycopersicum* L.	Tamatar	Yukpañcakam rasapūram
92	*Solanum nigrum* L.	Makoy	Yukpañcakam kṛṣṇaphalam
93	*Solanum virginianum L.*	Chhoti kateli	Yukpañcakam pītaprakaṇṭam
94	*Swertia chirayita* (Roxb.) H.Karst.	Chirayita	Ghūrṇobhadalakam tiktam
95	*Syzygium aromaticum* (L.) Merr. & L.M.Perry	Laung	Jambukaḥ lavaṅgaḥ
96	*Terminalia arjuna* (Roxb. ex DC.) Wight & Arn.	Arjun	Vīrakaḥ arjunaḥ
97	*Terminalia bellirica* (Gaertn.) Roxb.	Bahera	Vīrakaḥ akaṣaḥ
98	*Terminalia chebula* Retz.	Harad	Vīrakaḥ harītakaḥ
99	*Tinospora cordifolia* (Willd.) Hook.f. & Thomson	Giloy	Saptaśirikā aromapatrā
100	*Trachyspermum ammi* (L.) Sprague	Ajwain	Yavānikā ugragandhā
101	*Trigonella foenum-graecum* L.	Methi	Methikā gandhibījā
102	*Urtica ardens* Link	Bichu ghas	Alipatrakam sajātam
103	*Viola odorata* L.	Banafsha	Vanapsikā nīlapuṣpā
104	*Vitex negundo* L.	Nirgundi	Nirguṇḍikā madhupapriyā
105	*Vitis vinifera* L.	Angoor	Drākṣikā gostanī
106	*Withania somnifera* (L.) Dunal	Ashwagandha	Aśvagandhakaḥ svāpakaraḥ
107	*Zanthoxylum armatum DC.*	Tambru	Tumburukaḥ kaṇṭikāṇḍaḥ
108	*Zingiber officinale* Roscoe	Sunthi	Ārdrakam sitauṣṭham

ABBREVIATIONS

ADC	=	Automatic development chamber
AMD	=	Automatic multiple development chamber
Fig	=	Figure
HPLC	=	High pressure liquid chromatography
HPTLC	=	High performance thin layer chromatography
Mg	=	Milligram
mL	=	Milliliter
mm	=	Millimeter
MP	=	Mobile phase
MS	=	Mass spectroscopy
nm	=	Nanometer
Rf	=	Retardation factor
RPM	=	Rotation per minute
S	=	Standard
T	=	Test sample
TLC	=	Thin layer chromatography
TTC	=	Twin trough chamber
UV	=	Ultraviolet
V	=	Volume
µm	=	Micrometer
254	=	Short wave length UV light, reflection
366	=	Long wave length UV light, reflection

BIBLIOGRAPHY

[1] Balkrishna, A. *Ayurveda Jadi Booti Rahasya.,* **2014**; *1*: 1-467.

[2] Balkrishna, A. *Ayurveda Jadi Booti Rahasya.,* **2014**; *2*: 575-1077.

[3] Balkrishna, A. *Ayurveda Jadi Booti Rahasya.,* **2014**; *3*: 1097-1572.

[4] Poole, C.F. *Instrumental Thin-Layer Chromatography.,* **2015**; 31-105.

[5] Reich, E.; Schibli, A. *High-Performance Thin Layer Chromatography for the Analysis of Medicinal Plants.,* **2006**; 1-240.

[6] Wagner, H.; Bladt, S. *Plant Drug Analysis: A Thin Layer Chromatography Atlas.,* **1995**; *2*: 349-369.

[7] Wall, P.E. *Thin-layer Chromatography: A Modern Practical Approach.,* **2005**; 6-170.[http://dx.doi.org/10.1039/9781847552464]

[8] https://powo.science.kew.org

[9] https://www.camag.com

Acharya Balkrishna

SUBJECT INDEX

Acharya Balkrishna

I

J

K

L

M

N

O

V

W

Z